Albert Rössler

Design of Experiment for Coatings

Cover: Adler-Werk Lackfabrik GmbH & Co. KG

Albert Rössler
Design of Experiment for Coatings
Hanover: Vincentz Network, 2014
European Coatings Library
ISBN 3-86630-885-X
ISBN 978-3-86630-885-5

Please ask for our book catalogue
Vincentz Network, Plathnerstr. 4c, 30175 Hanover, Germany
T +49 511 9910-033, F +49 511 9910-029
books@european-coatings.com, www.european-coatings.com

Satz: Vincentz Network, Hanover, Germany
Druck: Quensen Druck + Verlag GmbH & Co. KG, Hildesheim, Germany

ISBN 3-86630-885-X
ISBN 978-3-86630-885-5

European Coatings Library

Albert Rössler

Design of Experiment for Coatings

Preface

For sure you have experience with such problems: the complex production process or the demanded coating properties are affected by several factors (grinding time, binder content, amount of additives, drying time, etc.). You would like to improve the system and do not know which are the most dominant factors. No wonder, as long as your formulation is assembled by 10 components. The experimental setup get out of hand, results are unintelligible and the documentation is fragmentary at the end. In this situation you might got already the advice to try the statistical approach by design of experiment (abbreviated DoE). Nice, but what is that? Beside, statistics sounds like mathematics and at the moment you are too busy for such things?

Well, time and some distance from the daily business will be necessary, if you read this book. However, the aim is a short and generally understandable overview regarding the effects und benefits of DoE by examples in step with actual practice. The examples are based on real-world applications in coatings formulation. This adds the important strong application flavour to this book, makes it useful as a reference tool and show that statistics doesn't have to be that dry and difficult mathematics. In addition, requirements for the approach are illustrated, which are quite often general principles in experimental design and project management. It is not possible to describe the topic without any statistics or mathematics, but formulas and derivations are at the least possible amount in this book. So, the prerequisites are relatively modest. In addition, there is no claim to be complete, because the topic and connected mathematic methods as well as tools for data analysis are too extensive. Due to this, the textbook familiarize first-time user and already practicing laboratory assistants, engineers and chemists with possible applications and high potential of the method. It is important to succeed the entrance and to get a feeling of success by try and error. It's no problem, if not everything is clear in the first run. Experience about useful applications and limits will be generated in the course of time. After that, this technique will attend you for sure as valuable providing tool for a long time. The skills acquired in dealing with this method can also be employed in other applications, such as adhesives and sealants as well as all other engineering and scientific issues.

Chapter 1 introduces the method and shows especially the difference to the classical approach. With an example directly out of the "coating-formulation kitchen" it will be helpfully shown, how the systematic method DoE can reduce the daily mania in the research laboratory. Derived from today's criteria of success in R&D, also general principles of project management and design of experiments are illuminated.

Chapter 2 reflects the design methods and introduce into the important preliminary step of drawing up a strategy of experimentation. State of the art experimental designs are presented together with their properties. This should help for precise selections later on in practice. Planning as preliminary step is the basic for all further details and the potential savings are naturally enormous. Therefore, planning is of much higher importance as the subsequent data analysis. A lot helps a lot in this context!

Chapter 3 is devoted to the data analysis and interpretation. During this period the basic principle „nothing ventured, nothing gained", can be applied. Only data from perfectly planned and designed experiments ends up in reasonable interpretations. On the other hand, the best set of data is useless, when no interpretation and good compression is performed. Where is the limit? How valid and precise are the conclusions? These are some questions, which will be answered in Chapter 3.

Chapter 4 introduces the standard optimization processes and describes how to convert modelling results into concrete action. The determination of cause and effect correlations during the identification of the most dominant factors is only one part of the story. It is also of high importance to find those operating conditions, which lead to a maximum, minimum or a certain value (or range) of the target response. In addition, just optimizing a problem will end up very often in results, which are sensitive against small variations in the factor settings due to disturbances. To look on the issue in its entirety, also strategies for robustness studies are described, leading to processes and products that perform better in the field due to variability reduction. In a manner of speaking: finding a needle in the haystack.

Chapter 5 describes some DoE software solutions on the market. It is not necessary to reinvent the wheel once more and to create all experimental designs by hand. Modern and good designed programs assist the practitioner in the number-crunching.

Short text elements including basic principles are included in all book chapters. They summarize general and essential messages regarding design of experiments and performing experiments. These elements do not replace the rest of the book as a kind of summary, because basic principles are not covering methodical details. Basic principles only intensify relevant messages regarding the philosophy of the DoE-concept. Therefore, they are not used inflationary.

Research and development in coating industry is highly affected by experience. If somebody is working long enough in this branch, a comprehensive treasure of raw material and formulation knowledge will be acquired. This experience is very important and is beside the application knowledge the working basis. However, research and development underlie today tough necessities, because innovative ideas, higher rates of success, shorter time to market as well as lower R&D-costs should be realized in parallel. Therefore, both, a deep understanding of the increasing international customer demands and economic mentality becomes more and more important in research laboratories. In addition, knowledge about the correlations – independent if the recipes, the production or the application process are in focus – is essential for creating innovative, stable, robust, cost-optimized and high-quality products very rapid today. Thus, methods which support purposefully and efficiently working are crucial to be successful.

Design of experiments is one possibility, to reply the challenges of today and to manage the important free space for creativity due to higher efficiency. This book can assist you in this challenging task and shows how coating formulators can benefit from DoE.

Innsbruck, Austria, Mai 2014

Albert Rössler

Contents

1 Design of experiments - systematic mania?

1.1 Design of experiments as part of the challenges and criteria of success in modern R&D

A rapid change engraves the world of business in the 21th century. Due to the progressive globalization and the connected increasing competition, market conditions become tougher. Companies are constrained to establish a restrictive culture of continuous product- and process-improvement. Otherwise quality problems and low efficiency end up in a dramatic financial loss. Only fast, flexible and powerful operating will survive in the future. Cost aspects, quality of the products and short cycles of development (time to market) [1, 34, 36] are the essential directions of impact for enterprises today (see Figure 1.1).

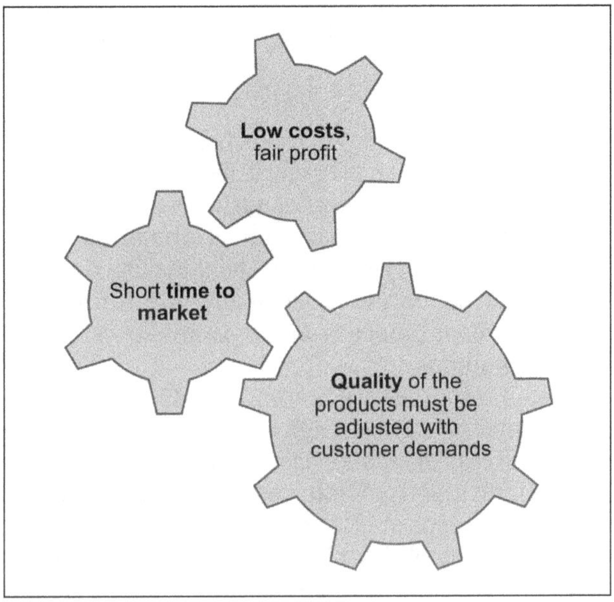

Figure 1.1: Criteria of market success [1]

Derived from these aspects, research and development (R&D) underlie today tough necessities, because innovative products, higher rates of success, shorter time to market as well as lower R&D-costs should be realized in parallel [2]. Thus, the following most essential **criteria of success in modern R&D** can be mentioned [1, 34, 36]:

- A deep understanding of the (increasing international) customer demands **(success factor customer value orientation)**.
- Goal-oriented and efficient advance during product and process development considering the principles of economics - maximum amount of relevant information with a minimal effort for experiments **(success factor efficiency)**.
- Knowledge about the correlations within the process chain - independent if the recipes, the production or the application process are in focus **(success factor system knowledge)**.
- Generation of additional synergy and velocity due to a platform strategy or a robust building block system already in the period of development [27, 35]. Due to that, it is not necessary to develop a completely new product for each customer demand (and to produce it later on) **(success factor efficient diversity, managing the product complexity)**.
- Integrated thinking as mentality already during research and development, to avoid a so called local limited island optimization without looking on the up- and downstream demands (e.g. fast and efficient production) **(success factor integrated mentality and efficiency)**.
- Stability and quality of the products and processes **(success factor quality)**.

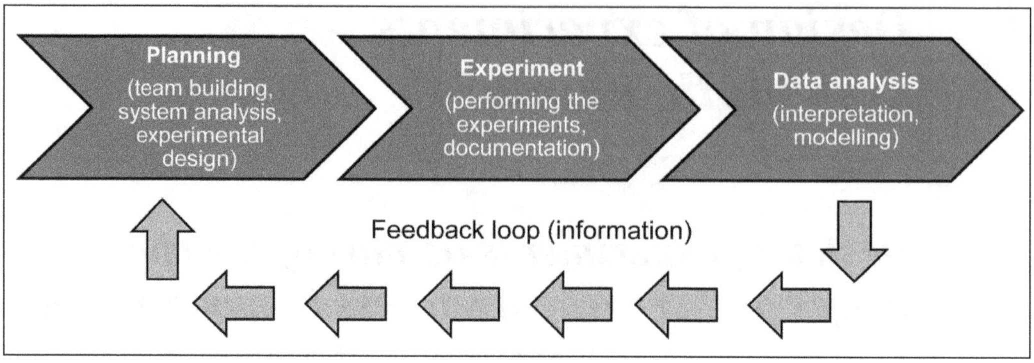

Figure 1.2: Steps during a R&D-project inclusive a feedback loop for the information

These criteria are only completely successful, if in addition

- the R&D department is based on a high performance organisation, raising the synergies between different R&D groups, technologies and issues as well as supporting dynamics in contact with the customer or during dealing with comprehensive issues due to a strong project organization in parallel to the structural organization **(success factor structure)** [31], an optimized project management inclusive a project culture is existing [32].
- the organisation and the company culture are the homeland of creativity and continuous improvement **(success factor innovation culture, discussion culture and failure management)**.
- enough resources are present to ensure the essential free space for creativity and short reaction times also at high capacity utilization **(success factor capacity planning)**,
- process cost calculation is in focus and manufacturing costs are always under consideration in their entirety **(success factor process thinking)**.
- R&D team consist of flexible, highly qualified, team-minded and motivated employees, who live this philosophy, share knowledge with each other, are open for new aspects – leading the company daily to success **(success factor employees)**.

Central element of each development in natural science and engineering is the **experiment**, including measurements or simulations on the computer (see Figure 1.2). It is an elementary unit of each research project. Unfortunately, in technical oriented branches it's still an illusion to develop a new product just on the writing desk. The interactions in systems (e.g. surfactants, thickener, binder, solvent) are too complex and simulation tools are very often imprecise (e.g. regarding the long term behaviour). The high degree of interactions in coating formulations (e.g. a white wall paint) is shown in Figure 1.3a). Already the fact, that 13 or even more different components are present in the system, which might be from 10 different suppliers, describes the high complexity of chemical as well as physical interactions. Interactions are also obvious due to the fact, that single components have a different active potential in different formulations. It is not enough to look only on single interactions and to test raw materials in standard formulations. An example based on the gloss and spread rate of white pigmented wall paints is shown in Figure 1.3b). Wetting agent A can achieve quite good results in binder 1, but not such a high level in binder 2. However, the behaviour of wetting agent B is the opposite way around. Thus, the quite common 1:1 exchange of raw materials has only limited explanatory power about the potential of single raw materials. **Integral formulation optimization** is the essential part of the game, to raise the whole potential. This ends up, at least in case of high-end products, inevitably in an intensive and open cooperation between coating manufacturer and the raw material supplier **(success factor network, cooperation, open innovation)**. Otherwise robust products cannot be realized in an acceptable time frame.

In addition, in practice mainly formulations with a long history are present and the path of formation development cannot be reconstructed. Very often components are added to a formulation due to some problems, but nobody will remove them afterwards - independent if the trigger for the problem is still present or not. Last but not least the extensive unknown chemical and physical design of the raw materials makes the situation even worse. Altogether, geniality and innovation consists still today - as already *Thomas A. Edison* said - of 1 % inspiration and 99 % transpiration, or in other words, no pain, no gain.

Partly, the experiments are also connected with an extreme high effort, a lot of time and high costs. Thus, methods which support the above mentioned aspects by

- a systematic, structured approach during (research) projects, experiments, etc.,
- a reproducible and objective analysis of cause-and-effect-correlations in case of experiments, products and processes and
- an adaption of the experimental effort according to the requirement during (research) projects

are crucial for the success of the companies.

Pos.	Component	%	Supplier
1	Water	7.9	
2	Surfactant	0.7	A
3	Defoamer	0.3	B
4	Cellulose	0.1	C
5	NaOH	0.2	D
6	Binder	40	E
7	Titanium-dioxide	10	F
8	Filler 1	4.2	G
9	Filler 2	4.2	G
10	Filler 3	8.2	H
11	Filler 4	23	H
12	Solvent	1.2	I
13	Thickener	0.1	J

Figure 1.3a: Integral formulation optimization: Complexity and interactions in a coatings formulation [40]

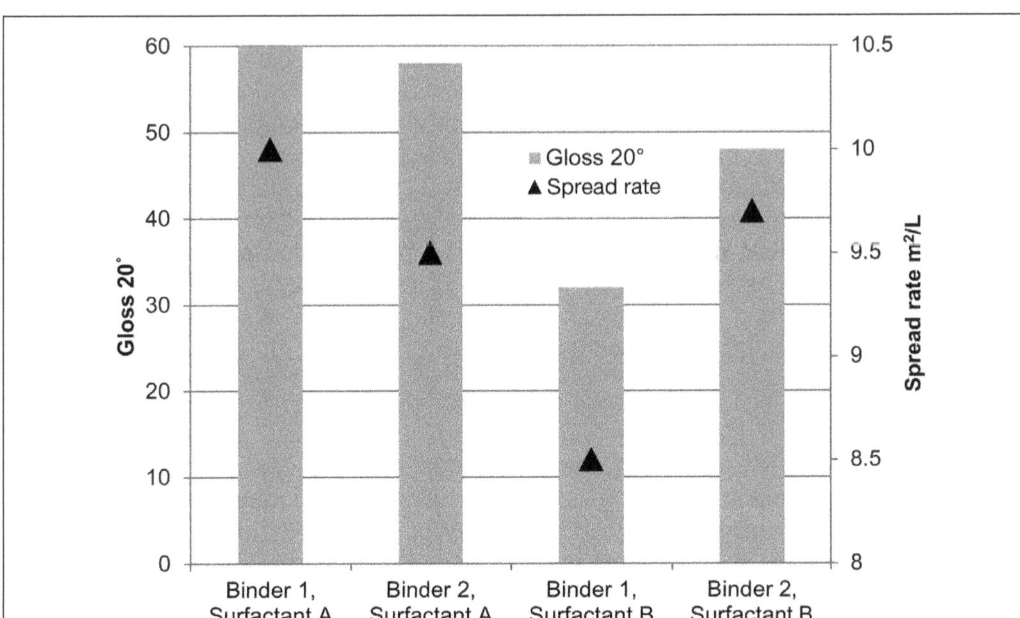

Figure 1.3b: Integral formulation optimization: Gloss and spread rate for a white wall paint as a function of the binder and the wetting agent [33]

Every approach which supports or even enables this method of operation is, because productive and value added, welcome. Only by this way [22, 23]

- unnecessary loops for variations during the development of products and processes can be reduced or even avoided,
- precise statements can be performed and subsequent steps can be initiated,
- the relevant influencing factors of a system can be identified **(success factor system knowledge)**,
- certain lacks in products or processes can be identified already preventive by a systematic analysis of influence and undefined or hardly controllable influence factors already in the development stage. This ensures a sustainable success with robust products and processes that can be used in a large application window.
- exact prediction of the experimental effort can be performed,
- all relevant experimental data are available in a reproducible manner after the experiments,
- not necessary experiments can be avoided,
- technicians will be faster in their daily business,
- the odds of success for realizing real technology jumps in an acceptable time frame with defensible costs are finally increasing.

The importance of increasing the system knowledge should not be underestimated **(success factor system knowledge)**. Of course each development is generating a comprehensive treasure of empirical values about the realized product, the raw materials, the applied aggregates, etc. However, thereby hardly any quantitative statements are generated. Outsider will have the feeling, that these empirical values are more than less only qualitative feelings of single persons with a low reproducibility. As a consequence, precise predictions about system behaviour cannot be realized. Unfortunately modern high performance products permit no tolerance concerning this matter. Designed as boosters for the company, a perfect balanced relation between cost structure and property profile is crucial. Simultaneously all functionalities of the products must be realized and continuously improved with a decreasingly number of raw materials suppliers. In addition, the performance characteristics should be robust against disturbances. Only an extensive and for everybody in the company available, documented factual knowledge about the relevant parameters of the system enables – by combining with the empirical values and intuition – efficiently procedures for the development of new products or for the fast elimination of troubles. In parallel, these skills allow a significant improve in the performance of the coating systems, because the whole complex and interacting system is optimized. System knowledge is also essential for the successful application of the already emphasized standardized technological platforms **(success factor efficient diversity, managing the product complexity)**, because the building blocks will be used in many systems [27]. Utilizing rigorous standardization will reduce variation and create flexibility and predictable outcomes. Thus, it is important to have a good knowledge about these modules.

Overall, these methods lead to a cost reduction, optimized time schedules and dramatic reduction in time to market (money) for new products – finally representing a modern, powerful R&D-strategy. Especially useful auxiliaries join in the phase of planning, which represents the basic for each project and each experiment (see Figure 1.2) [1] **(success factor project management and front-loading)**. In this period the strategy for further experiments is defined. Errors from this period influence the whole project and can compromise the aim of the project. In addition, the costs per each change in the product design or production process are increasing drastic during the project time (see Figure 1.4) [3]. A rule of thumb tells, that the cost for bug fixing decuple from each project step to the next [19]. To describe this also with an example, think about planning a house. In the very beginning, it has to be decided if wood or bricks are used for construction. Later on, maybe if the house is already under construction, any chance from wood to bricks will be very expensive. In the worst case, the house has to be pulled down. A good planning increases the odds of success!

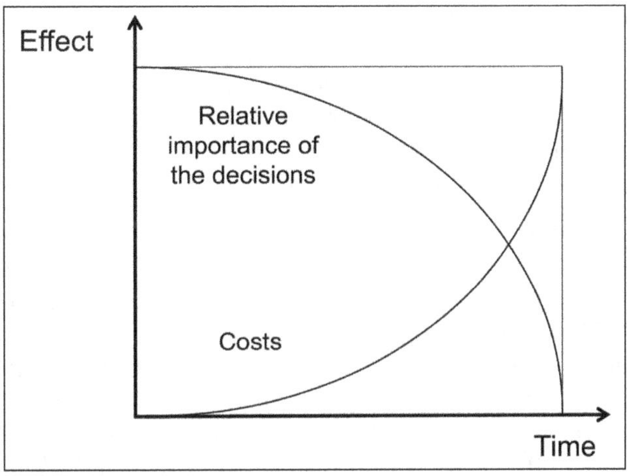

Figure 1.4: Influence of decisions on the target achievement and caused costs during the progress of a project [3]

▶ ***Basic principle:*** Planning represents the basic step for each project. A rule of thumb tells, that the cost for bug fixing decuple from each project step to the next, because errors from the very beginning influence the whole project and can compromise the aim of the project. Therefore, front loading in the development process is a basic principle, because there is a maximum design space to explore alternative solutions thoroughly.

All people are planning, when they would like to do something, which is different from the daily routine. If somebody would like to buy a car, he will think first of all which demands should be fulfilled: A certain label? A diesel engine or an alternative fuel vehicle? Five doors to insert the buggy into the car comfortable or a small and compact car to find a parking place simply in the city? Afterwards a sequence has to be defined to reach the aim (car dealer, newspaper, etc.) [30].

Also during experiments one or more factors are changed depending on the question behind. Thus, experiments are different from observations! Due to this fact, each experimental series need – as buying a car – a **strategy of experimentation**. The **experimental design** is aligned with the **aim/target** at the very beginning and is afterwards executed step by step. However, in the daily practice of a laboratory the conception of experiments is carried out – in contrary to buying a car – very often not highly reproducible and neatly arranged. This is for sure not always necessary, because of dealing with simple questions and a huge treasure of empirical values. However, it is not supposed to derive each experiment of a series by random or by intuition just due to the result of the previous experiment (see also random-method and intuitive method in Chapter 1.4.1). This procedure is not calculable in costs and not reproducible for outsiders. Only planning can prevent generating not necessary experimental data and ensure that all relevant information is present to verify the statements.

▶ ***Basic principle:*** Each experimental series has to be based on a strategy. An experimental design is defined at the beginning aligned with the desired goal and is afterwards executed step by step. Each experiment of a series cannot be derived by random or by intuition based on the result of the previous experiment. This procedure would be not calculable in costs and not reproducible for outsiders.

Planning is therefore more than just facing problems in a certain way. First of all, the target has to be defined as precise as possible [30, 32]. What is the problem and what is the desired result are the most important questions in this period. It is crucial, that there is a clear indication by simple, quantitative and easily measureable (by a validated method) **definition**

of the target (see also Chapter 1.6 and 2.1). Target definition is team work. If there is no clear aim, each participant has its own individual target. Stress, conflicts and inefficiency are pre-programmed. Very often, people are subject of false conclusions by thinking that everybody in the team has the same picture in mind. Ask 6 persons to draw a mode of transportation and you will get 6 solutions from airplanes up to cars. The target agreement was not precise enough. Without such a target agreement, projects will end up in a catastrophe.

The worst case will happen, if the target is moving during the project. This should be prevented, but keep in mind that planning is always a hypothesis which has to be adjusted with cumulative knowledge. The thesis can be summed up by a statement of the German field marshal *Helmuth von Moltke*, that "no plan survives contact with the enemy". Demands of customers can change abrupt, e.g. when the competitor comes to the market with new attractions or trends. Therefore, projects should be started only, when enough capacity is available. Otherwise the risk of moving targets is increasing due to slow progress and lag of time in the project [28, 29]. Be aware of the fast technical development in some branches!

▶ ***Basic principle:*** Defining the purpose of a project is crucial for the ability of the project team to solve the issue. Due to this, the team can focus on the relevant factors and targets [28]. Teams must have a clear indication and understanding of the target, which should be declared in written form, simple, quantitative and easily measureable by a validated method.

Experimental series should generate objective and authoritative information (based on data). It is therefore - as shown in Figure 1.2 - partly necessary to have a feedback loop. This means, that the design is modified due to some new insights from data analysis and further experiments are done. Thus data are never equal to information. Methods, which can also assist during this interaction between design and results of experiments, are of outstanding value. Experiments are usually iterative and especially in case of first efforts or learning experiments the degree of feedback is quite high.

▶ ***Basic principle:*** Experiments are usually iterative. Thus, the design of experiments is very commonly modified due to new insights from data analysis of previous experiments. This feedback is essential to have an optimized progress in the ongoing project. The degree of feedback is increasing with decreasing system knowledge at the beginning of the experiments.

For the sake of completeness it should be added, that targets based on facts does not lead without fail to **product efficiency**. Clever developments are for sure the core of R&D (lat. *ingenium* - cleverness). Unfortunately, it is very often a perfect stimulation for technicians and engineers, to break a technical record. Therefore, it is important to keep always in mind, that product properties just fulfil the customer demands **(success factor customer orientation)**. It is not necessary - or even more precise useless - to create more features, than aligned with the customer **(principle of lean development** to avoid muda - *muda, japanese* meaning uselessness; waste; wastefulness) [37, 38]. Benchmarking and value analysis [20, 21] are perfect possibilities to check this issue. In general, the products should be as simple as possible, but not simpler. The question, which functionalities can be omitted is of the same importance as one about the demanded skills [28]. Companies will be only successful in the long term and with sustainability, when the products are as good as necessary, but not as good as possible. Already an old Chinese saying told us, that the worm must taste the fish, not the fisherman.

Systematic approaches in experimental design represent one way to reply the actual mentioned challenges of R&D. Permitted targets, which are focused by experimental data collection and subsequent evaluation, can only be reached in an efficient way by such a systematic method.

▶ ***Basic principle:*** Basic idea and purpose in experimental design is to improve the efficiency (quality and quantitative amount) in product and process development by eliminating the intuitive "trial and error" or even "random" methods with a more systematical approach, transparent correlations and a minimal effort for experiments (which is known already before starting). Thus chances at the market can be perceived earlier by a substantial reduction of the development time.

Economic aspects are today an integral content in the mind of natural scientists and engineers. Design of experiment (DoE) can also be understood as philosophy or mindset, which is as package highly cross-linked with supplemental up- and downstream tools like project management, quality management, benchmarking, Quality Function Development (QFD), Failure Mode and Effects Analysis (FMEA - see Appendix 8), combinatorial material research [24], platform strategy [27], etc. This philosophy can revolutionize product and process design in many kinds. Design of experiment is also an integral element of quality planning and quality management in R&D, construction and process planning [22, 23]. Of all the statistical tools used today, design of experiments offers the most power for making breakthroughs for variability reduction in coatings formulation and processing. Finally, this element perfectly suits to the principles of lean-development, which bear on the analysis of structures, processes and tools regarding dissipation and the definition of possible countermeasures [37, 38].

In the last years laboratory automation has become a really hot topic in coating science. High-throughput-sites supply more and more the development of coatings [25, 26]. Is this increasing the efficiency? Such instruments can only be used in a meaningful way, as long as design of experiments is done at the front end and data analysis is adjusted on this method. Then, the potential of a lab robot can be developed to a whole extend. Intuitive experiments, just looking what happens, rough fathom of limits, etc. has to be done before in the classical laboratory. Otherwise, the effort for programming the machine is too high and the overall equipment effectiveness (OEE) too low. With a combined working strategy High-throughput methods can be enhanced to High-output-methods. Finally market chances appreciate earlier and at better conditions.

1.2 A typical experiment in coatings formulation

The roman philosopher *Lucius Annaeus Seneca* already mentioned, that "the way through education is very long, but short and effective by examples". Thus, let us consider as introduction a typical experiment in coatings formulation: adjusting the surfactant content on the amount of solids like pigments, fillers, etc. Precise, we focus on the development of a waterborne filler concentrate made of barium sulphate by using a polymer based dispersant agent. Aim is to adjust the viscosity at a certain value (approximately 2000 mPas at a shear rate of 100 s^{-1}, 20 °C) and to realize in parallel the maximal possible transparency. From earlier experiments it is known, that the maximal amount of an inorganic thickener is limited due to its negative influence on other coating properties to a constant value of 0,3 %. A variation of the concentration or using other thickeners (e.g. polyacrylic base or cellulose ethers) is also not possible. What are you doing?

By intuition, the surfactant will be premixed with water and afterwards the filler will be added by dispersing. A starting point regarding the filler and the surfactant content is known from experience or can be found in a reference formulation. However, before starting a lot of questions have to be answered:

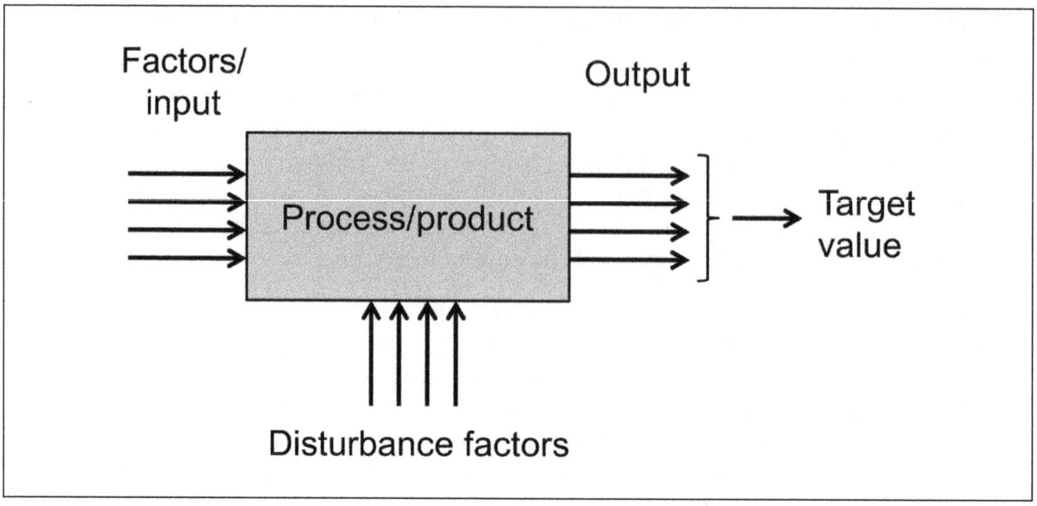

Figure 1.5: General description of a technical system

- What is the optimal relation of the raw materials, which is necessary for a viscosity close to the target value of 2000 mPas (at a shear rate of 100 s^{-1}, 20 °C)?
- What will be the maximal possible transparency?
- Is there a composition with a viscosity in the target area and with a maximal transparency?
- Is the composition changing, when in addition, the constancy of viscosity during 8 weeks temperature storage is included as target value?
- What will be a robust process window regarding viscosity variations?
- Is there an influence of the dispersing time or other parameters (e.g. sequence of adding the raw materials, temperature)?
- etc.

Before we will answer these questions in detail, some terms and definitions should be clarified.

1.3 Factors, Levels, etc. - some vocabulary at the beginning

All questions from the example in Chapter 1.2 about the content of surfactant and transparency of a filler concentrate can be generalized in the following way (see also Figure 1.5):

- **Target value or response** is transparency and viscosity. The target values are always derived from measureable values. Sometimes there is only one response, very often even more.
- **Influencing factors** are the content of surfactant and of filler. These two parameters may have an influence on the response by changing their level.
- Random is unfortunately (or sometimes also fortunately) always a companion of our experiments. Some variables are often undefined and not controllable. Maybe laboratory assistant A is adding the surfactant faster than laboratory assistant B, maybe temperature is increasing during an experimental series, maybe batch A of the raw material is different from batch B, etc. Such imponderables are called **disturbances or uncontrollable factors.**
- All disturbances together, including environmental effects and all not controlled factors are called **noise factors.**

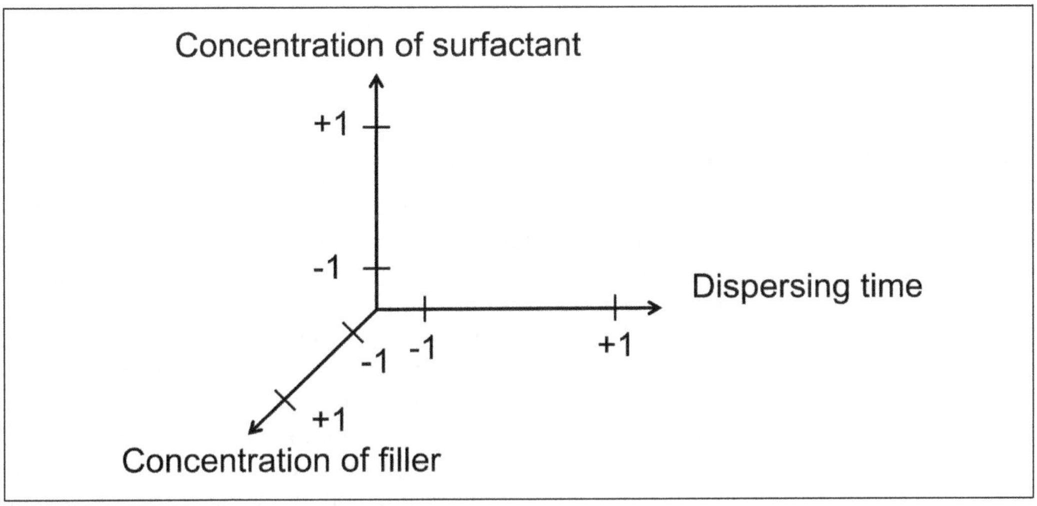

Figure 1.6: Example with two levels for three factors

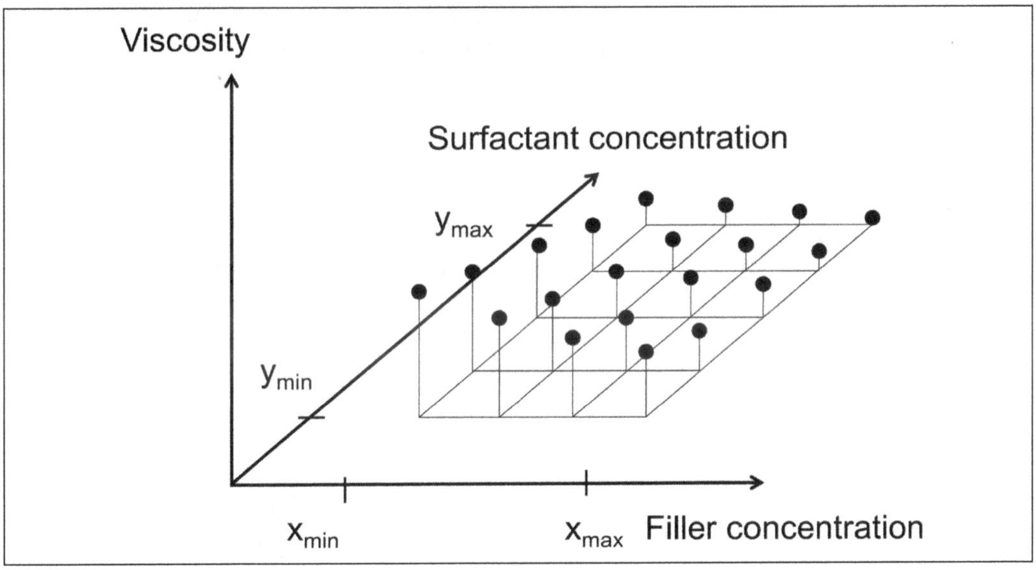

Figure 1.7: Example with four (solid content) and five (content of surfactant) levels for two factors regarding the response viscosity

To answer the questions from Chapter 1.2, a design of experiment is necessary. Experiments always need resources like manpower, instruments, etc. and time to market or even better money must be as short as possible. Thus, there will be soon a conflict, between accuracy and precision of the results and the effort which is necessary for this level. The **number of trials per each level and the number of levels** at which a factor is analyzed is important for the **total amount of experiments/runs.** In case of the filler concentrate the solid content can be fixed at 20 and 40 % to check the effect on transparency and viscosity. Naturally also three levels at 20, 30 and 40 % would be possible. However, design of experiment is very often done at two levels – high and low. It is common to call these levels "low" and "high" and denote them by "+" and "-" or "+1" and "-1", respectively. This is especially important in the screening phase to minimize the amount of experiments, because this method still allows identifying the significant factors. By increasing the number of levels a network can be created in the experimental area (see Figure 1.6 and 1.7).

1.4 Classical design of experiments and the limitations

1.4.1 Conventional methods – more diversity is not possible

Considering the example from Chapter 1.2 what would be the best approach to find out the optimized ratio of ingredients regarding the target value of 2000 mPas viscosity (at a shear rate of 100 s^{-1}, 20 °C)? Usually, the solid content will be varied from a starting point (e.g. 20 %) at a fixed value of surfactant concentration (e.g. 5 %) until there is no further improvement. Afterwards the optimized content of the filler will not be changed during optimizing the surfactant concentration. The other possibility is the inverted procedure by optimizing the surfactant concentration in the first step and keeping the content of the filler at a constant level and a variation of the filler content afterwards at the optimal level of the surfactant concentration. In both cases, the focus (variation and optimization) is always only on one factor after the other. The experiments will end up from a starting point of 5 % surfactant and 20 % filler with an optimized recipe of 30 % filler and 5 % surfactant (see Figure 1.8).

This is the so-called classical experimental design or **one-factor-at-a-time method** (see Figure 1.8). The approach was developed already in the 16th Century by the British natural scientist and philosopher *Francis Bacon*. For a systematic analysis of the correlations each factor (e.g. content of surfactant) will be varied and all other variable factors are held constant. This is in accordance to the **Ceteris paribus principle** (lat. *Ceteris paribus* – all other things being equal or held constant) to guarantee a simple and unique allocation of an effect to one variable. In a first step, the first factor is varied starting from a certain point and all other factors are constant. The variation takes place until there is no further improvement or even a worse result is achieved. After a move back to the best setup until now, the variation starts once more by varying the next parameter. The optimized adjustment of the first factor is not changed during the following experiments. Mostly the adjustment of the factor which was held constant in the first series (surfactant) will change and won't be constant as in

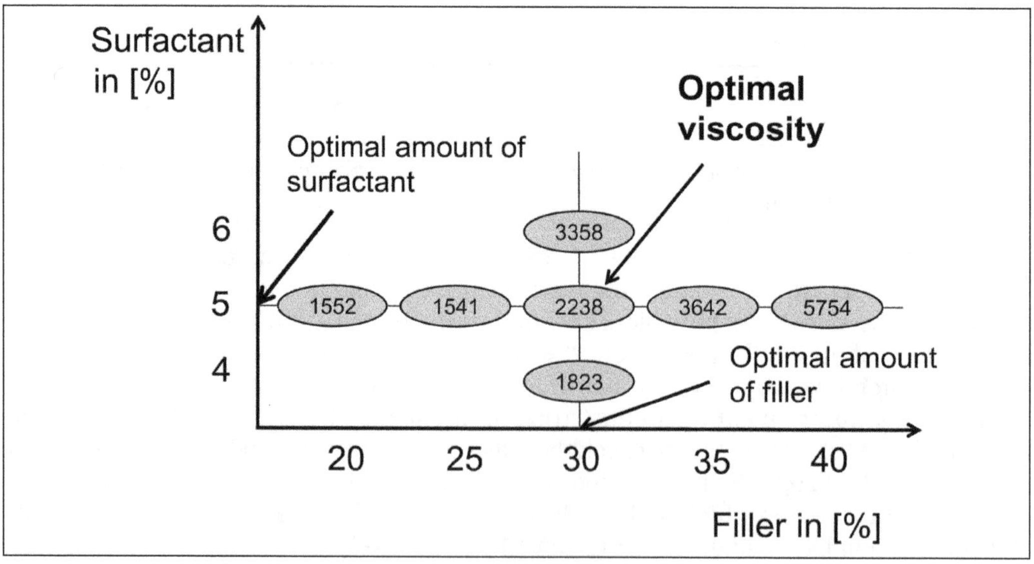

Figure 1.8: One-factor-at-a-time method for the example with the filler concentrate. The target viscosity (numbers in the data points in mPas) can be achieved in the experimental area with 30 % filler and 5 % surfactant.

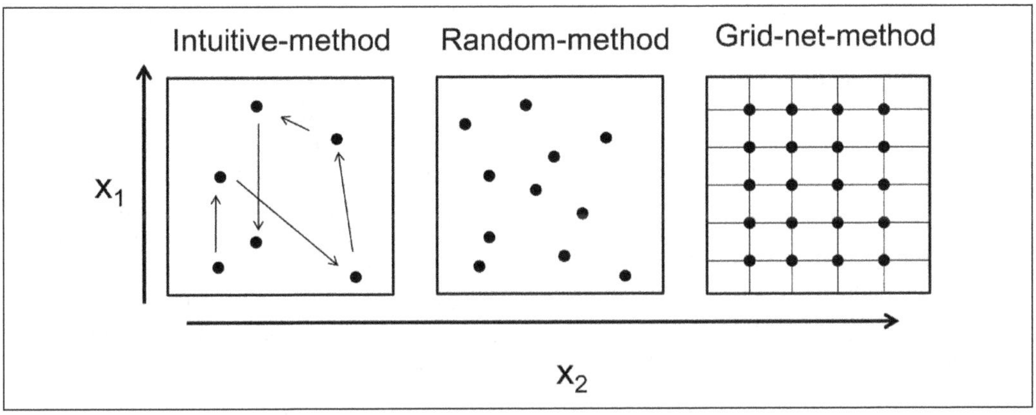

Figure 1.9: Some other conventional methods of experimental design

Figure 1.8. It is more a fortune coincidence, that the starting adjustment represents already the optimized dimension.

In principle there are three further and highly oppositional methods of experimental design: **intuitive method, random method** and **grid design method** (see Figure 1.9). The simple variation of one factor (e.g. as concentration series) is not considered in this overview. In case of the already in Chapter 1.1 mentioned and very often applied intuitive procedure a starting point is chosen depending on the experience of the person doing the experiment. Based on the results of the first experiment, the fixation of the next experimental setup takes place. This act is continued until achieving the target value. Assuming high experience about the system under evaluation, this method leads very soon to success. Unfortunately, most of the experiments are not reproducible, because the complex assumptions are not sufficient documented. The planning and the estimation of cost are also impossible. However, intuition is not bad by itself, notably it is essential when experience is combined with knowledge. In cases, where only fragmentary knowledge is available, or contradictory opinions are present during a decision procedure, intuition is essential. The same is valid in cases, when the rational mind is at its limit due to a high degree of complexity. The second method is randomizing the factor levels in the experimental area. Thus, reproducibility, systematic data analysis and cost calculation are even worse. The third method produces a grid in the experimental area and analyses the setup with a high intensity. However, this is not the statistical approach in experimental design as often asserted. Method two and three need really a lot of experiments and are not so present. The classical approach is the one-factor-at-a-time method and the intuitive method.

1.4.2 Limits in case of the classical approach

1.4.2.1 Number of experiments in case of many factors

In general the number of runs can be calculated by Equation 1.1 from the number of factors and levels (assuming that there is always the same number of levels for each factor).

Equation 1.1: Number of runs = Number of factors · (Number of levels − 1) + 1

How can we derive the equation? Looking on a system with two factors and two levels per each factor (e.g. low and high level of surfactant like 1 and 4 % or 20 and 40 % filler), we will end up with an experimental design shown in Table 1.1. Three experiments are necessary to execute the experimental program. In case of three factors already four experiments (see Table 1.2 and Figure 1.10) are necessary. Every additional factor (e.g. dispersing time) with

Table 1.1: One factor experimental design with two parameters and two levels

Number of experiments	Factor	
	X_1	X_2
1	-	-
2	+	-
3	-	+

Table 1.2: One factor experimental design with three parameters and two levels

Number of experiments	Factors		
	X_1	X_2	X_3
1	-	-	-
2	+	-	-
3	-	+	-
4	-	-	+

two levels ends up with one additional experiment (look on the dark grey point in Figure 1.10). Thus, the number of experiments is equal to the product of the number of factors and the number of levels. Regarding the number of levels, the experiment in the central point has to be subtracted (see Figure 1.10). This point is for all factors the same and without subtracting multiple counting would happen. This point will be considered as its own experiment and has to be added in the equation at the end.

For many factors the experimental effort is extensive, because the number of data points is very high with this method. This is reasonable, considering the high number of not useful experiments and levels per each factor. Otherwise it is impossible to find the position of the optimum. These additional experiments are not considered in Equation 1.1, because the equation assumes always the same number of levels. In practice this assumption is rarely grantable. To ensure an adequate statistical reliability the huge number of replicates has to be performed (see Figure 1.25 and Chapter 1.6). Therefore, the one factor method is not economic for more than two factors.

▶ ***Basic principle:*** The one-factor-at-a-time method and the intuitive method are always less efficient than other methods due to the huge amount of experiments for more than two factors.

During the further book design of experiments will be demonstrated as highly effective method, which can realize an acceptable extent of experiments despite the simultaneously variation of all factors and a clear allocation of the effects also with many factors (see Chapter 2.2 and Figure 1.25).

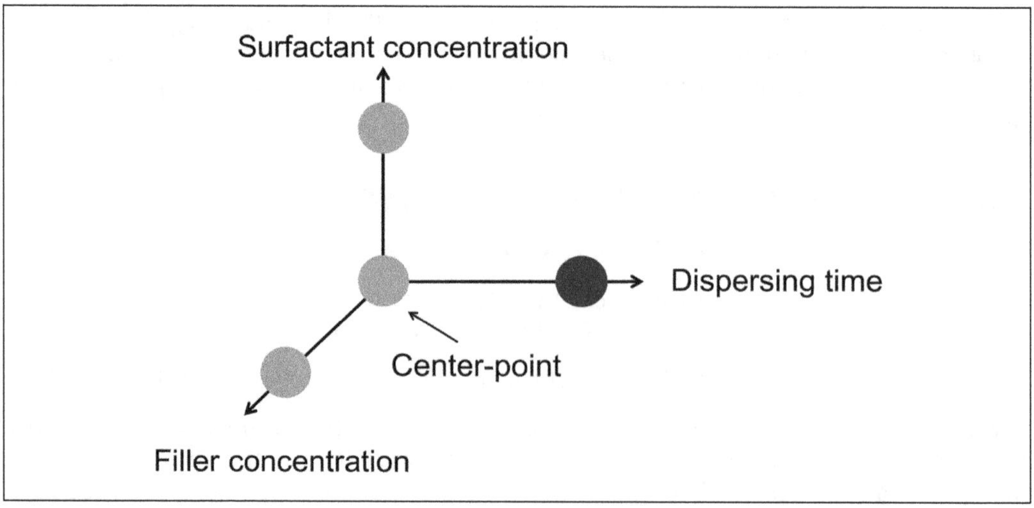

Figure 1.10: One factor experimental design with two (light points) and three (additional point in dark grey) with two levels

- ***Basic principle:*** Not every difference between two values is an effect. Every measurement is imperfect and scientists must judge frequently whether a numerical difference is a manifestation of the indeterminated errors inevitable in all measurements. All statements have to take into account the unavoidable fluctuations of experimental data (replications of the measurement, calculation of the mean and the confidence interval). The difference between two values is significant, when the averages separated by at least two times of the value of the confidence interval. However, not every statistically significant difference between two values is large enough to have any practical value. This has to be proofed in a second step.

- ***Basic principle:*** Significance is in general relevant for any target formulation or the quality control of products. Experimental results seldom agree with just one value. Thus, tolerance limits of the product specifications has to be defined. In addition, the variance of the measuring method should consume only a certain amount, ideally only 20 %, of the specified tolerance limit. Otherwise, there is no free space for other fluctuations like from the production process, e.g. due to quality fluctuations in case of raw materials. In addition, a precision is faked which can never be fulfilled: A factory can never produce more precise than measure the specifications.

The approach to analyze data by confidence intervals is also valid for the one-factor-at-a-time method. However, the necessary replications of measurements have different influence depending on the experimental design. If it is necessary to assure a measurement by 4 replicated experiments, in case of the one-factor-at-a-time method $3 \cdot 4 = 12$ experimental runs has to be done (see Figure 1.24a). With every additional factor the number of runs is increasing by 4. In case of the DoE-approach only 8 runs are necessary, because the design is replicated twice and in each data point two measured values are available (see Figure 1.24a). Nevertheless at each level of each factor 4 experiments exist. Thus the average main effects are even with a lower number of experimental runs as precise as those from the single-factor experiment. If a ratio of the number of experimental runs of both methods is calculated, the DoE-approach is 12/8 = 1.5 times more efficient than the one-factor-at-a-time method (see Figure 1.25). In case of three fac-

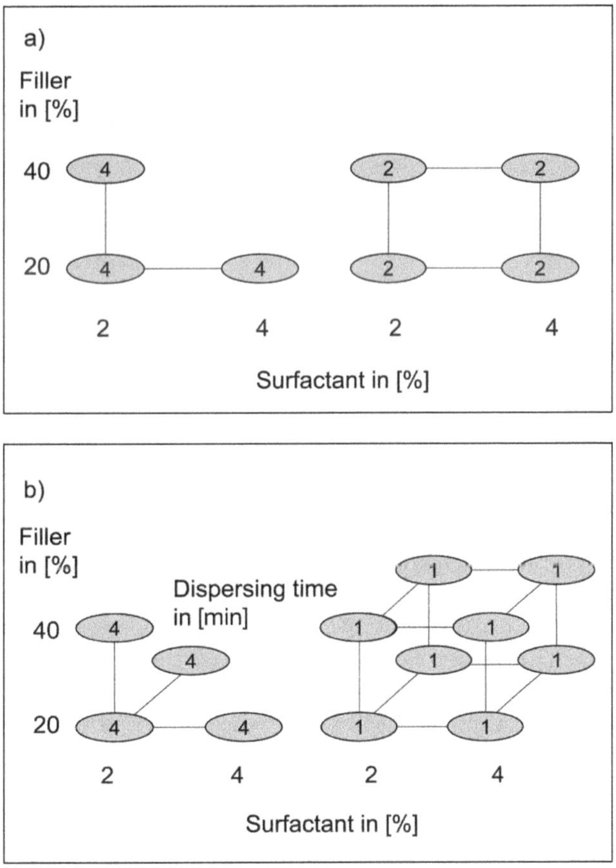

Figure 1.24: Comparison of the one-factor-at-a-time method (always at the left side) with the design of experiments approach (always at the right side). The values in the data points represent the number of necessary experiments (multiple experiments): a) two factors b) three factors

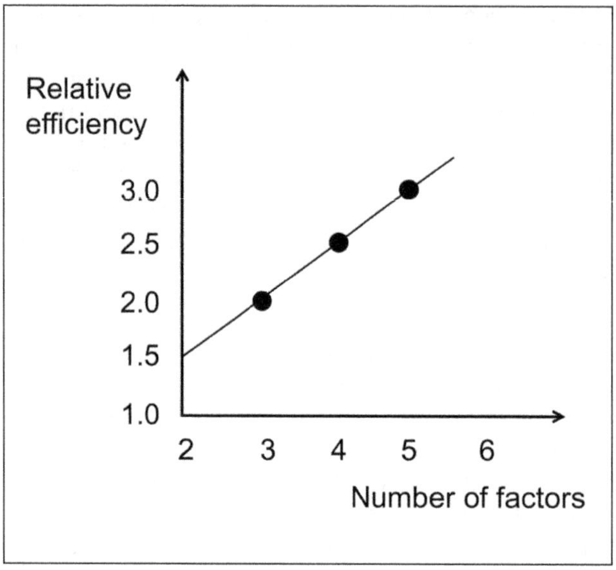

Figure 1.25: Relative efficiency of design of experiments vs. one-factor-at-a-time method[13]

tors the ratio is 16/8 = 2, because in case of the one-factor-at-a-time method 16 experimental runs are necessary and in case of DoE only one run per each data point (in sum 8 experiments) has to be done (see Figure 1.24b and 1.25). In this case at each level of each factor 4 pairs of values are available and the statistical validation is equal. This calculation can be done for any number of factors and is shown in Figure 1.25.

▶ ***Basic principle:*** With regard to experiments, which always need to be carried out multiple times, the statistical DoE approach is much more efficient than the one-factor-at-a-time method.

1.7 Models – pictures of reality

Each experiment provides some results. Once collected, these data are investigated using data analysis, because in most cases correlations between the data are assumed. This cause and effect correlations, relating the changes in the factors to the changes in the responses, enable a conscious control of the system and an extraction of in-depth information. To make a statement about this issue, we need a (mathematical) model, which represents the basic mechanisms [8]. It enables also afterwards a precise prediction about measured data or an optimization. Main effects and interactions are very often not enough meaningful.

Models are like pictures for us. They are not reality, but approximate representations of some important aspects of reality. A classic example is the landscape or an electric toy train. Think about atomistic models. In this case the word is already included in the definition. Models can be derived from well-established theoretical fundamental laws of science (like the Ohm's law or the free fall), or have only an empirical background. In the second case, which is much more important in practice (because the system is too complex), correlations between input and output (like viscosity as a function of filler and surfactant content) elucidate how factors influence the response of a system work. Design of experiments produces always an **empirical model**, which illustrates on a quantitative base the correlation between factors and the response. This is an essential advantage regarding the simple collection of data points. To know the influence of the pH-value of the binder on the effect of the thickener is much more valuable than only the knowledge about the certain value of the optimized pH condition. This compression can never give a clear explanation of the basic mechanisms behind the issue. The system is like a **black box** and a description of the correlation between input and output takes place (see Figure 1.26). But how is it possible to determine the correlation of the system?

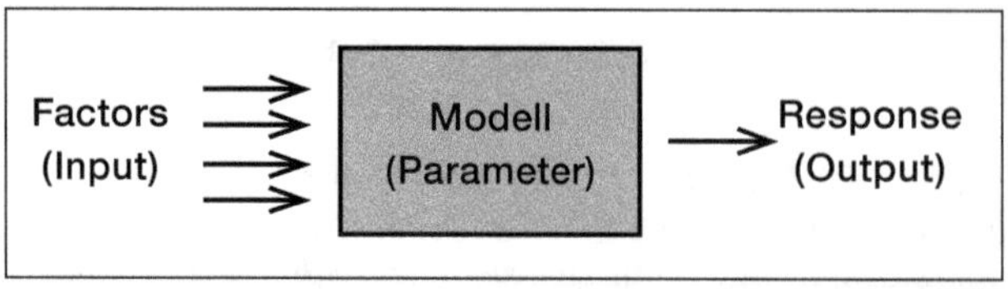

Figure 1.26: Empirical general Black Box-Model of a process or system

If all data are available, a straight line or even a curve can be drawn through the cloud of data points by a lineal. This line characterizes the correlation between the factor and the response and is called „best fit straight line“. To work with this correlation much more efficient and to perform also predictions, usually it is tried to describe this line with a mathematical equation.

From the perspective of mathematics, the simplest model is the **linear model** with the so called **equation of a line** (see Figure 1.27 and Equation 1.10). In case of one factor, the equation is:

Equation 1.10: $y = a_1 \cdot x + a_0$

y ... response
x ... factor
a_1 ... slope of the line
a_0 ... axis intercept

In case of more factors, the model is extended to:

Equation 1.11: $y = a_0 + a_1 \cdot x_1 + a_2 \cdot x_2 + a_3 \cdot x_3 + \cdots + a_n \cdot x_n$

with
y = response
x_1 to x_n = factors
a_1 to a_n = slope of the line respective model parameters regarding each factor
a_0 = intercept of the line

The determination of the model equation can be done by the so called **regression analysis** (see Chapter 3.3 and Appendix 7). This tool is offered by many software packages including EXCEL.

Main effects and interactions can be used to describe the effect of the factors on the response. Thus, it is obvious, that they can also be used to derive the linear model. But how can this be done?

The main effect is the average difference between the response at the high (+1) and the low (-1) level of the effect. The distance between the two levels is in this case 2. Thus, the slope is given by half of the main effect (see Figure 1.28). The intercept a_0 can be calculated from the average of all data points. Therefore, design of experiments offers the possibility to calculate directly the model parameters (at least for a two factorial design (see Chapter 3.3)).

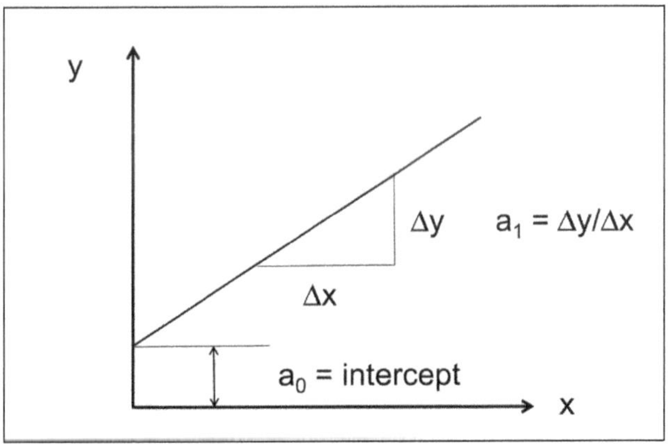

Figure 1.27: Schematics of the equation of line

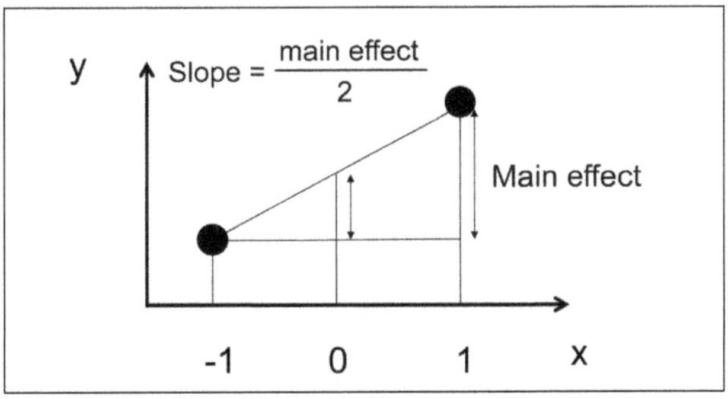

Figure 1.28: Determination of the model parameter from the main effect

Linear models can only describe easy and not complex systems. There is no degree of freedom. 1 + 1 is in a linear model always 2. However, sometimes the whole is more than the sum of each part (interactions, non-linear behaviour, see Chapter 1.4.2.2). Regarding the mathematical description in experimental design three very often used types of models can be distinguished (described for two factors, see Figure 1.29).

Equation 1.12: $y = a_0 + a_1 \cdot x_1 + a_2 \cdot x_2$ *(linear model)*

Equation 1.13: $y = a_0 + a_1 \cdot x_1 + a_2 \cdot x_2 + a_3 \cdot x_1 \cdot x_2$ *(linear model with interaction)*

Equation 1.14: $y = a_0 + a_1 \cdot x_1 + a_2 \cdot x_2 + a_3 \cdot x_1 \cdot x_2 + a_4 \cdot x_1^2 + a_5 \cdot x_2^2$ *(quadratic non-linear model)*

In case of two-factorial designs these models can also be derived directly from the DoE-approach.

▶ ***Basic principle:*** Design of experiments offers in contrast to the classical approach (for both, one-factor-at-a-time or intuitive method) the calculation of an empirical model. The model describes the system with a mathematical approach. Real effects are estimated by the model coefficients. This opens up the possibility to optimize the system and generate predictions about certain parameter settings by simulation. Thus the system knowledge is increasing **(sucess factor system knowledge)**. However, models are not reality, but approximate representations of some important aspects of reality.

Looking once more on the example of the filler concentrate, the linear model including an interaction of Equation 1.15 can be derived from the main effects and the interaction (Table 1.6). The intercept a_0 is calculated from the average value of all data points, which are registered in Table 1.4.

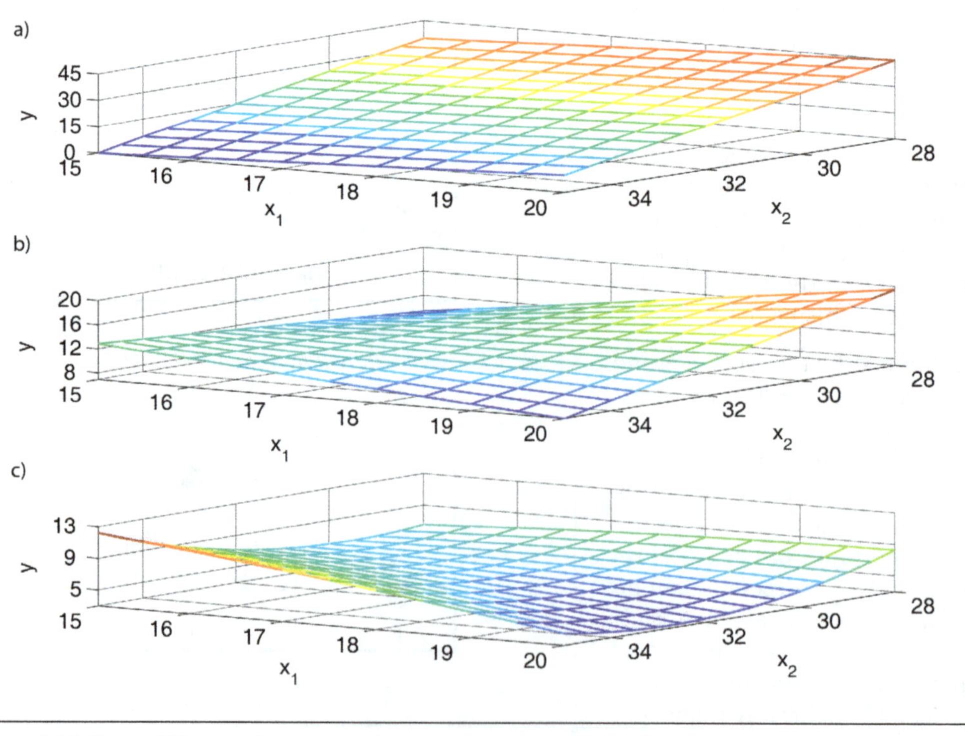

Figure 1.29: Three different relevant models in experimental design: a) linear model based on Equation 1.12, b) linear model with interaction based on Equation 1.13, c) non-linear model based on Equation 1.14

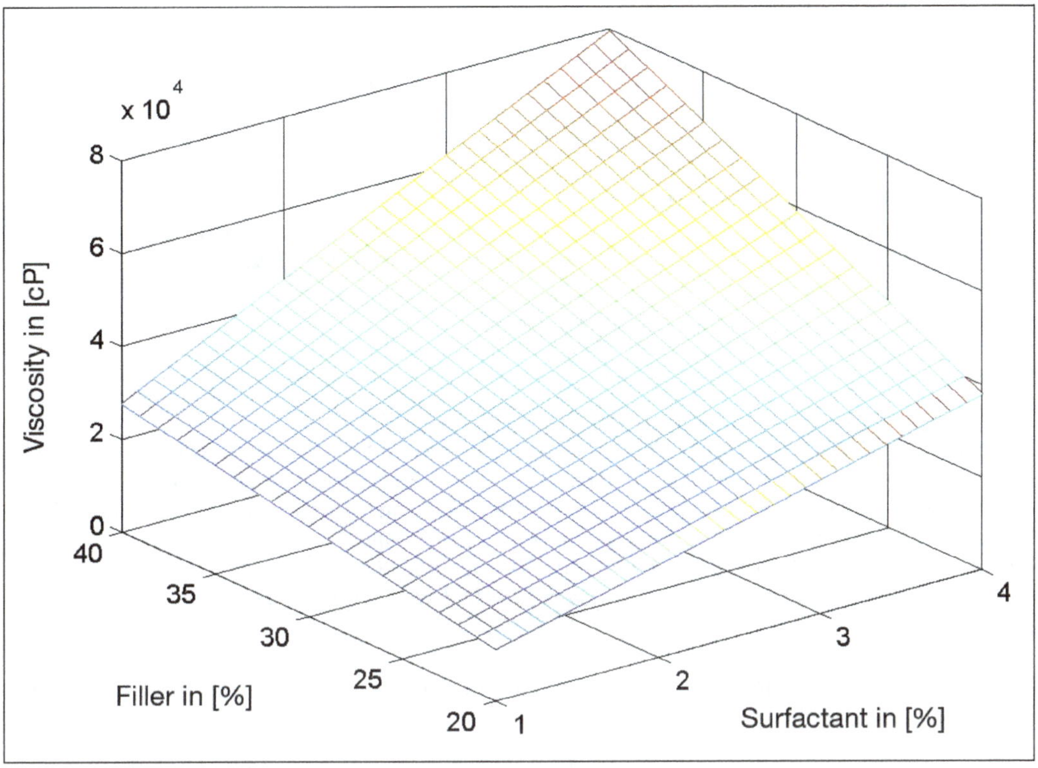

Figure 1.30: Response surface based on a linear model for the example of the filler concentrate

Equation 1.15: Viscosity (in mPas) $= a_0 + a_1 \cdot x_1 + a_2 \cdot x_2 + a_3 \cdot x_1 \cdot x_2$

$$= 4558\ \text{mPas} - 1373\ \frac{\text{mPas}}{\%} \cdot \text{Surfactant (in \%)}$$

$$+ 2928\ \frac{\text{mPas}}{\%} \cdot \text{Filler (in \%)} - 643\ \frac{\text{mPas}}{\%^2}$$

$$\cdot\ \text{Surfactant (in \%)} \cdot \text{Filler (in \%)}$$

The model is shown in Figure 1.30, perfectly illustrating the principle of interaction (see Chapter 1.5.2). The slope on both edges is completely different. This means that an increase of the surfactant concentration at the lower level of the filler content has a significant other effect as at the higher concentration level of the filler. An extreme example is shown in Figure 1.29b. In both Figures 1.29 and 1.30 systems are illustrates three-dimensional as so called **response surfaces** to see the impact of inputs on the response. A two-dimensional

Table 1.6: Calculation of the model equation directly from the main effects and the interactions for the example of a filler concentrate

Identification	Identification in the model	Main effect and interaction	Model parameter (= Main effect/2)
Main effect (Surfactant)	a_1	-2745 mPas	-1372.5 mPas
Main effect (Filler)	a_2	5855 mPas	2927.5 mPas
Interaction (Filler · Surfactant)	a_3	-1285 mPas	-642.5 mPas

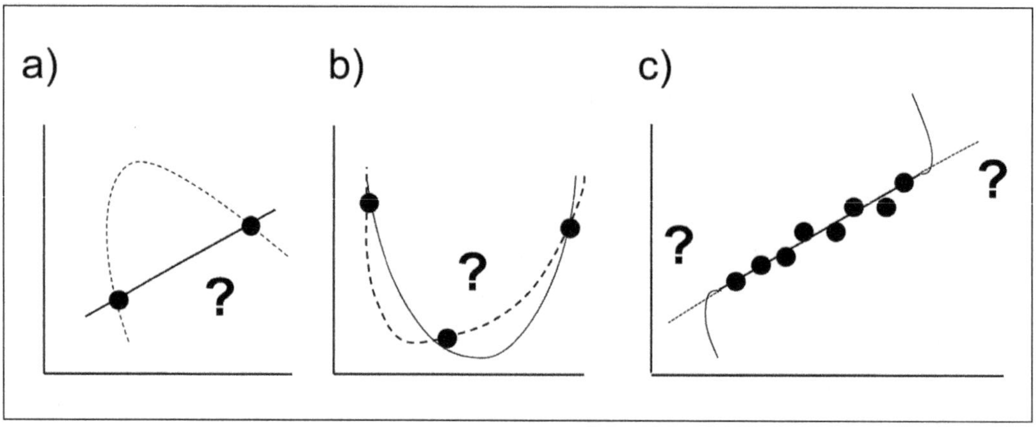

Figure 1.31: a) It is always possible to draw a straight line through two data points. Unfortunately it is not clear if there is in practice another function (dotted line); b) The Minimum of a response may be shifted to the wrong place, when analyzing a quadratic model only with three data points (dotted line), because the quadratic function is always symmetric to the optimum. Also in case of saturation curves very often a non-realistic maximum is indicated; c) Extrapolation of a model is forbidden. Without any further data points it cannot be clarified, which model is relevant beside the actual limits of the experimental area.

plot or map **(contour plot)** is also used very often and illustrates the response by contour lines (see Figure 2.16).

One picture due to fitting a model is worth a thousand numbers. Beside the well-arranged overview, Figure 1.29 shows also the much deeper insight about the correlations and mechanisms of the system in the whole experimental area. In addition, due to this step it is possible to clarify important further questions like

- the optimal recipe regarding viscosity (see Figure 1.32) and transparency,
- the range of operation conditions which can be expected as highly robust against viscosity fluctuations – see Chapter 4.3.

In principle the chosen generated design of experiments and the chosen model are intimately linked. It has to be taken into account that a two-factorial plan analyses only two data points (low and high level). Unfortunately through two data points only a linear model with a straight line can be drawn, because another model is not clearly definable. With three data points it is possible to create a non-linear model (see Figure 1.31). However, be careful with very simple models based on two or three data points (see Figure 1.31a and b). There is a high chance to run the risk of overseeing non-linear behaviour.

Models have always only a local explanatory power. Inside the analyzed area interpretations are possible. An extrapolation, meaning a prediction of values outside the analyzed area is forbidden (see Figure 1.31c and 3.6).

Every model has to be analyzed on its **plausibility**. Later on systematic methods will be shown about this topic (see Chapter 3). Especially additional data points show the explanatory power of the model quite good. Some basic principles of modelling are summarized in Chapter 3.3.3.

All previous statements did not take the fluctuation of the data into account. It was assumed that the model represents reality. However, in practice there is a certain fluctuation due to disturbances, errors, etc. on the response. Ultimately, also the model parameters have an uncertainty. Repeating the experimental design for the same conditions, each time ones

Typical fields of application are:

- Time effects in case of extensive experimental series (e.g. week 1 and week 2, etc.)
- The raw material of one batch is not enough for all experiments. Thus, a second batch has to be used.
- Personal exchange (e.g. due to shift work)

The formation of blocks is done in front of the experiments. Subsequent data analysis takes place (see Figure 2.1). Differences between the blocks will be studied by the block variable. An experimental design with blocks will be described in Chapter 2.2.5 and 2.3.2.

▶ ***Basic principle:*** Blocking the experiments into synchronized groups is a useful principle to combat and control sources of unwanted variability in measured data. This is especially valid in case of influencing factors which are not directly connected with the target value (different lab personal, batches of raw material, etc.).

Randomization and blocking are conflicting to a certain amount. Whereas blocking assigns data in a systematic way, randomization demands a random assignment. To overcome this contradiction, blocks are assigned in a systematic way, but inside the blocks data points are randomly positioned.

2.1.4 Normalization, centring, orthogonal design

In design of experiments, all variables are varied simultaneously. Usually a clear identification of effects is impossible in such a case. However, if the design of the procedure is well arranged, it is possible to take this hurdle. The secret behind is the so called **orthogonality**. An experimental design is always orthogonal, if all levels of the influencing factors are independent from each other. Let us select the data points where one factor is at the level +1: In these data points all other factors are of the same amount at level +1 and -1. Thus, their influence is compensated during calculation of the average value for the +1-level of the subject matter. The same is true for the level -1. Due to this, the average effect of the influencing factor under closer examination can be calculated independently from the levels of all other influencing factors [2].

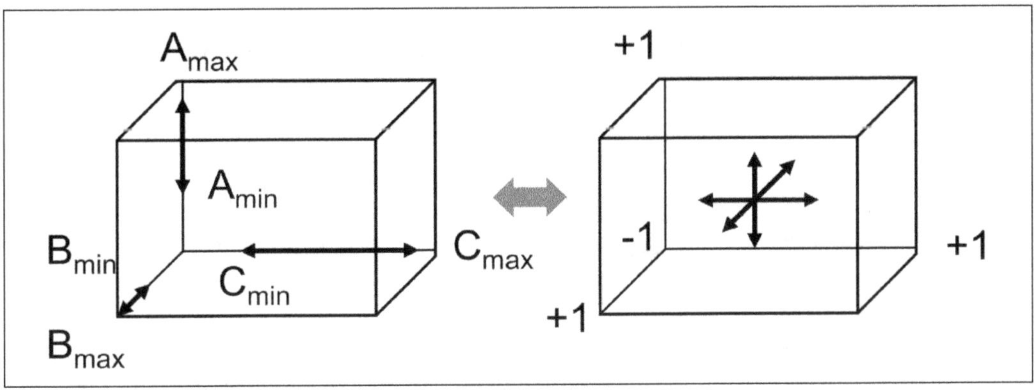

Figure 2.2: Normalizing and centring of the experimental area

In order to evaluate the influence of one factor in comparison to other factors, it is important to do a so called **normalization and centring** (see Figure 2.2 and 2.3). After this transformation all values are on a **coded scale** between -1 and +1. The translation of parameters into normalized and centred parameters can be performed by Equation 2.1:

Equation 2.1: $$A_{Norm} = \left(A - \frac{A_{Max} + A_{Min}}{2} \right) / \frac{A_{Max} - A_{Min}}{2}$$

The translation the other way around is done by Equation 2.2:

Equation 2.2: $$A = \frac{A_{Max} - A_{Min}}{2} * A_{Norm} + \frac{A_{Max} + A_{Min}}{2}$$

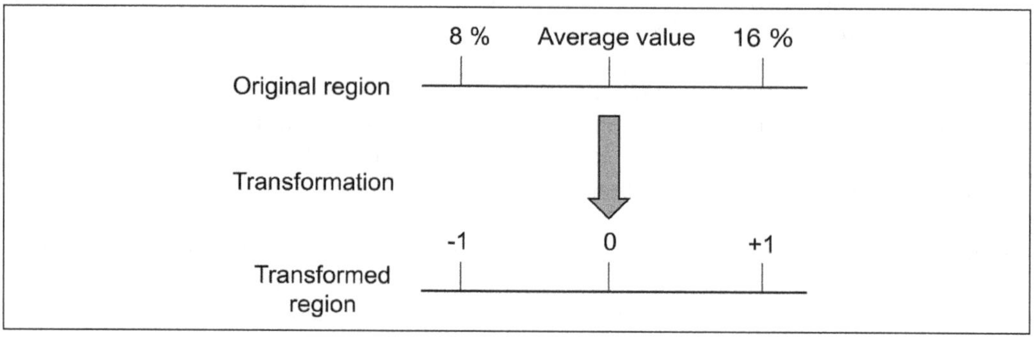

Figure 2.3: Example for the translation of data

2.1.5 Not realizable, irregular combinations - the experimental region

As described in the previous chapters, an experimental design should be orthogonal and, therefore, good suited for the analysis step. Thus, all planned experiments should be performed. However, sometimes certain planned levels cannot be realized (e.g. humidity can be adjusted manual only in steps of ±1 %, but the experimental design makes steps of 0.5 % necessary). In this case, these small deviations have to be taken into account in the analysis step to avoid distortion of the results. Preliminary pilot experiments are helpful, to get familiarized with the experimental area. Very often, the area can be reduced, or adapted to certain conditions so that the region becomes regular. However, sometimes technical aspects are against such changes. In addition, systems are present, where certain areas cannot be realized (e.g. phase change, explosive mixture, solubility limits, highly toxic by-products, and so on). In these cases orthogonalization gets lost, yielding **irregular experimental** regions. Therefore, the experimental area should be split into two independent parts, or at least as a replacement some experiments should be done on the **boarder of the experimental area** (see Figure 2.4).

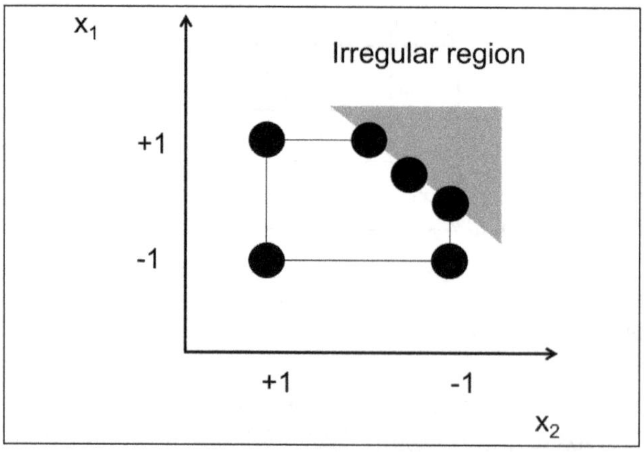

Figure 2.4: Implementation of data-points as a replacement in case of non-realizable setups and disallowed regions

As alternative to standard experimental design it is possible to make use of so called D-optimal designs (see Chapter 2.4.5). In these designs certain irregular regions can be excluded from the setup (so called **constraints** define disallowed or excluded regions by linear functions).

2.2 Factorial designs - the heart of DoE

Factorial designs are the central tool in design of experiments. The building principle is independent from the number of influencing factors. Designs include always all possible combinations of each single setup.

2.2.1 Two levels, two factors - 2^2-design

In the introductory Chapter 1.5.1 a simple two-level design for two factors has been already considered. In case of two variables with two levels - high and low (e.g. low and high amount of surfactant or rather 2 and 4 % surfactant or 20 and 40 % filler) - 4 combinations are possible (see design Table 2.3 and Figure 2.5). Analysis can be done as described in Chapter 1.5.1 by main effects and interactions taking the significance of the effects into account as described in Chapter 1.6.

Table 2.3: Design matrix for all experimental combinations in case of two factors and two levels

Number of experiments	Label	Factors	
		Surfactant, S	Filler, F
1	(1)	-	-
2	a	+	-
3	b	-	+
4	ab	+	+

Alternatively this analysis step can also be performed in form of a table very quick by hand (see Table 2.4). The experimental design is described in a table as already known. For each factor columns are drawn and filled with the sign of the level. These levels result directly from the experimental design. The columns of the interactions can be calculated by multiplication of the signs in the interacting factor column, e.g. AB = A · B.

The respective sign column is multiplied with the measured value to calculate the main effects and interactions by summation of each column and deviation of the sum by the number of experiments at high level or low level respectively (average value of the single effects). This approach can be derived from the current evaluation (see Equation 2.3 and 1.6).

Equation 2.3:

$$\text{HE(Surfactant)} = \frac{E(N,F-) + E(N,F+)}{2} = \frac{(a - 1) + (ab - b)}{2} =$$

$$= 1/2\ ((ab - b) + (a - (1))) = 1/2\ (a - (1) + ab - b)$$

Table 2.4: Table of signs for the calculation of the effects and interactions

Number	Surfactant	Filler	Surfactant · Filler	Measured value	Sign · measured value (Surfactant)	Sign · measured value (Filler)	Sign · measured value (Surfactant, Filler)
	Sign column						
1 – (1)	-1	-1	1	2360	-2360	-2360	2360
2 – a	1	-1	-1	900	900	-900	-900
3 – b	-1	1	-1	9500	-9500	9500	-9500
4 – ab	1	1	1	5470	5470	5470	5470
				Sum	-5490	11710	-2570
Main effect and interaction				Sum/2	-2745	5855	-1285

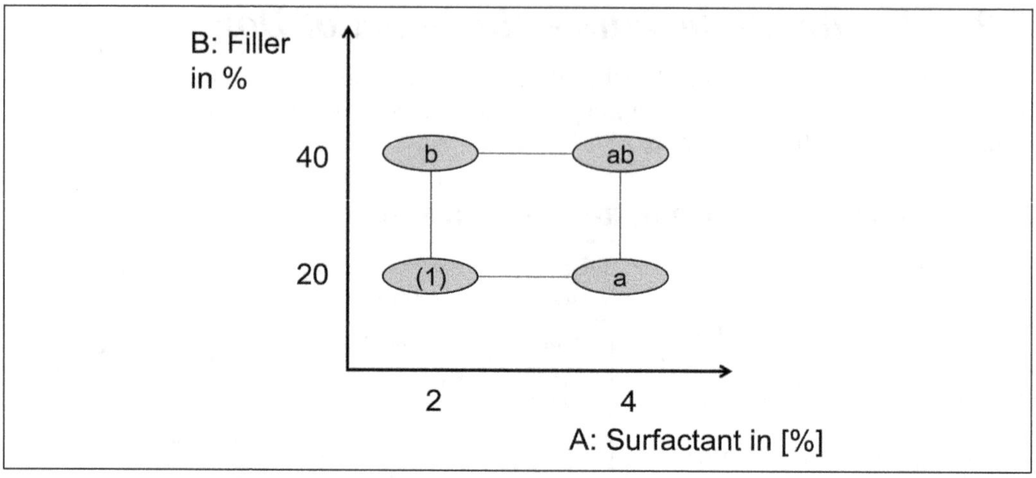

Figure 2.5: Design matrix and graphical illustration of the possible combinations with a factorial 2^2-design

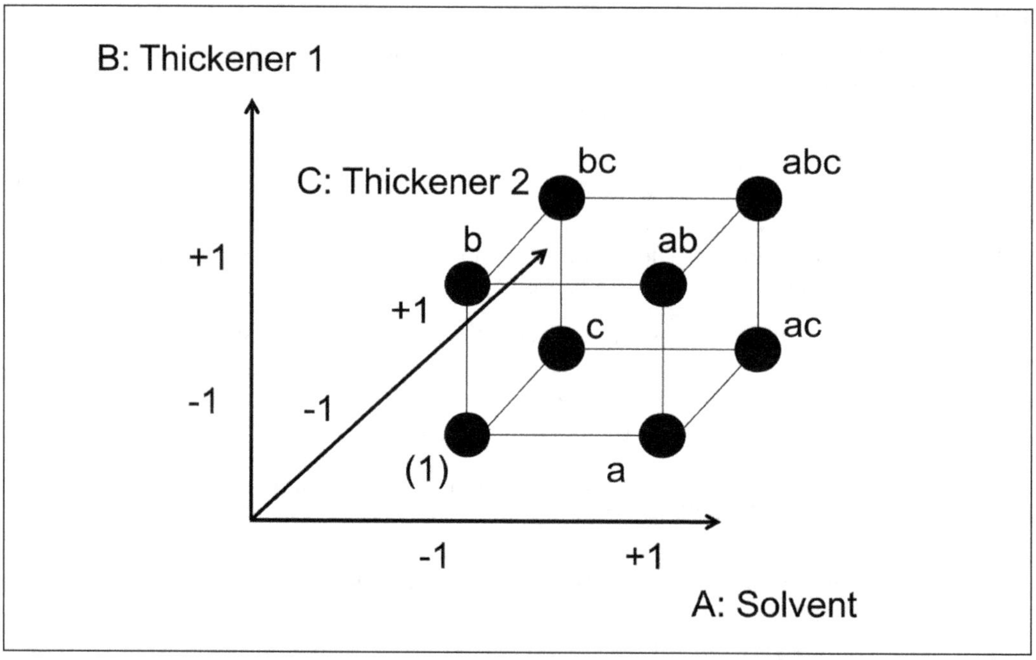

Figure 2.6: Experimental matrix and graphical illustration of the possible combinations from a factorial 2^3-design in case of the oscillatory-rheological measurement with an aqueous pigmented coating

The labels (1), a, b, ab represent certain data points. The current sign within the brackets in Equation 2.3 result directly from the experimental design respectively the sign column in the table of analysis (see Table 2.4; compare in Equation 2.3 the signs with those of the surfactant column). This procedure supports especially a possible analysis in spreadsheet programs, but also all other software based approaches.

2.2.2 Two levels, three factors - 2^3-design

In case of three variables already 8 experimental combinations are possible (see Table 2.5). These points are positioned on the edges of a cube (see Figure 2.6). The evaluation of this experimental design with main effects and interactions is in principle the same as

Table 2.5: Combination of experiments with three factors and two levels

Number of experiments	Factors		
	A	B	C
1 – (1)	-	-	-
2 – a	+	-	-
3 – b	-	+	-
4 – ab	+	+	-
5 – c	-	-	+
6 – ac	+	-	+
7 – bc	-	+	+
8 – abc	+	+	+

Table 2.6: Factor settings for the oscillatory-rheological test of an aqueous pigmented coating

Name	Low limit in [%] (level -1)	High limit in [%] (level +1)
Solvent, L	0	0.6
Thickener 1, V1	0.27	0.3
Thickener 2, V2	3.15	3.5

in case of the 2^2-design. For each edge of the cube single effects can be calculated. The main effect for one factor can be calculated by formation of the average from 4 single effects. Thus, a 2^3-design is assembled from 2 pieces of 2^2-design. Naturally the evaluation of a 2^3-design also can be done by tables. However, to get a better knowledge about the issue in this chapter the full evaluation is described step by step with an example.

Example: Influencing factors on the structural viscosity of an aqueous pigmented coating

The influence of two thickener (V1, V2) and the solvent (L) on the structural viscosity is evaluated in case of an aqueous pigmented coating. The target value is the ratio of the elastic modulus (G`) to the viscous modulus (G``) from an oscillatory-rheological measurement.

Properties of the factorial design with two levels (see Table 2.7):

- Number of experimental factors: 3 (see Table 2.6)
- Number of blocks: 1
- Number of target values: 1
- Number of experiments: 8
- Randomized: no

The average target value increase of a factor is called **main effect** HE. First of all, average values from all results at the higher and lower level have to be calculated. Afterwards the difference between these average values end up in the main effect (see Equation 2.4). As alternative, you can get the signs also out of the evaluation by tables (see Table 2.7).

Table 2.7: Design matrix for the oscillatory-rheological test with an aqueous pigmented coating

Number of experiment	Label	L	V1	V2	G`/G``
1	(1)	-1	-1	-1	1.17
2	a	1	-1	-1	1.12
3	b	-1	1	-1	1.14
4	ab	1	1	-1	1.25
5	c	-1	-1	1	1.14
6	ac	1	-1	1	1.22
7	bc	-1	1	1	1.24
8	abc	1	1	1	1.43

Equation 2.4:

$$HE(V1) = \frac{ab + b + abc + bc}{4} - \frac{(1) + a + ac + c}{4} = \frac{1.25 + 1.14 + 1.43 + 1.24}{4} - \frac{1.17 + 1.12 + 1.22 + 1.14}{4} = \frac{1}{4}(-(1)-a + b + ab - c - ac + bc + abc) = \frac{1}{4}(-1.17 - 1.12 + 1.14 + 1.25 - 1.14 - 1.22 + 1.24 + 1.43) = 0.103$$

Also in this case it is obvious, that the experimental setup already produces the signs inside the bracket (compare the column general term in Table 2.7 with the column thickener 1, V1). The main effects from thickener 2 and solvent are computed in the same way.

Equation 2.5:

$$HE(V2) = \frac{abc + bc + ac + c}{4} - \frac{(1) + a + b + ab}{4} = \frac{1.43 + 1.24 + 1.22 + 1.14}{4} - \frac{1.17 + 1.12 + 1.14 + 1.25}{4} = 0.09$$

Equation 2.6:

$$HE(L) = \frac{a + ab + abc + ac}{4} - \frac{(1) + b + bc + c}{4} = \frac{1.12 + 1.25 + 1.43 + 1.22}{4} - \frac{1.17 + 1.14 + 1.24 + 1.14}{4} = 0.08$$

Interactions from every single 2^2-design, which is part of the 2^3-design are calculated as already shown in case of the 2^2-design in Equation 1.9 by formation of the difference between the single effects. Exemplarily the calculation of the interaction between the two thickener WW(V1V2) is described at both levels of the solvent (Equation 2.7 and 2.8).

Equation 2.7:

$$WW_{L-}(V1,V2) = \frac{E(V1,V2+,L-) - E(V1,V2-,L-)}{2} = \frac{(1.24 - 1.14) - (1.14 - 1.17)}{2} = \frac{0.13}{2} = 0.065$$

Equation 2.8:

$$WW_{L+}(V1,V2) = \frac{E(V1,V2+,L+) - E(V1,V2-,L+)}{2} = \frac{(1.43 - 1.22) - (1.25 - 1.12)}{2} = \frac{0.08}{2} = 0.04$$

The total value of the so-called two-factor interaction WW(V1V2) results from the average value of both single interactions at the high L+ and low level L-.

Equation 2.9:

$$WW(V1,V2) = \frac{WW_{L-}(V1,V2) + WW_{L+}(V1,V2)}{2} = \frac{1/2\,(E(V1,V2+,L-) - E(V1,V2-,L-)) + 1/2\,(E(V1,V2+,L+) - E(V1,V2-,L+))}{2} = \frac{0.065 + 0.04}{2} = 0.05$$

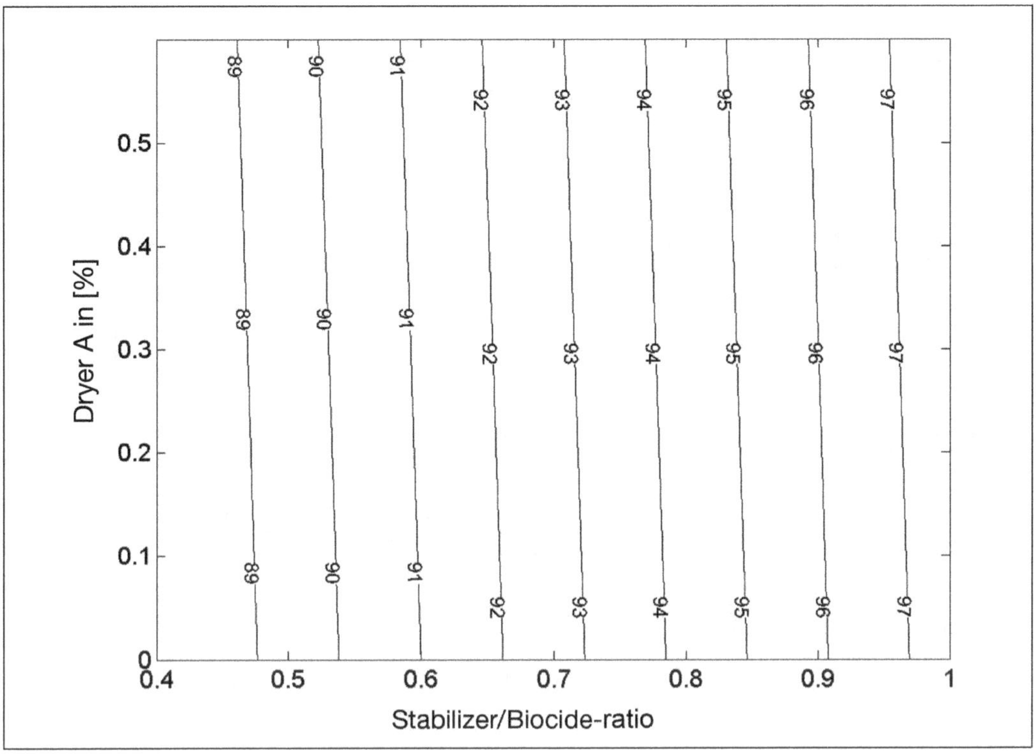

Figure 2.16: Contour-plot with consideration of blocking for the example recovery rate of a biocide. The plot is based on dryer B = 0.2 % and block +1

If blocking is not considered as independent variable, no significant influence of all parameters can be detected (Figure 2.15b). All effects are smaller than the error bars and not significant. The effect of all parameters is smaller than the noise generated by the difference between the used raw material batches.

An illustration of the block effect is also shown in Figure 2.1. The sensitivity of all experiments is increased and the confidence interval drastically reduced, because blocks with conditions as equal as possible (same raw material batch) are formed. If it is not distinguished between the two raw material batches, the data set is mixed up and a considerable higher variance arise. This is already the case, because the two blocks are at a different level (see Figure 2.1b). In this example the uncertainty of the effects, indicated by the error bars, are given by the confidence intervals of the regression model, which was used to compute the contour-plot. A replication of the runs was not possible. Hence, the error-bar just indicates the goodness of the regression and the quality of the experimental design (see also Chapter 3.3).

▶ ***Basic principle:*** The validity of statements and predictions is improved due to the consideration of noise factors by blocking. There is no demand for a higher experimental effort using fractional factorial design.

2.3.3 Types of fractional factorial designs

In principle several types of fractional factorial designs are distinguished by the **resolution** (in roman numerals) based on the number of independent variables. A design of low resolution experiences complicated confounding of effects.

Table 2.20: Maximal resolution of fractional factorial designs - the type is given by the roman numerals [3]

Number of experiments	Number of factors							
	3	4	5	6	7	8	9	10
4	2^{3-1} III							
8	2^3 Full factorial	2^{4-1} IV	2^{5-2} III	2^{6-3} III	2^{7-4} III			
16		2^4 Full factorial	2^{5-1} V	2^{6-2} IV	2^{7-3} IV	2^{8-4} IV		
32			2^5 Full factorial	2^{6-1} VI	2^{7-2} IV	2^{8-3} IV	2^{9-4} IV	2^{10-5} IV
64				2^6 Full factorial	2^{7-1} VII	2^{8-2} V	2^{9-3} IV	2^{10-4} IV
128					2^7 Full factorial	2^{8-1} VIII	2^{9-2} VI	2^{10-3} V

A few important types are:

- **Type III**: Main effects are not confounded by themselves, but confounded with two-factor interactions. This resolution is critical, because main effects might be distorted or reversed without any advice.
- **Type IV**: Main effects are not confounded with two-factor interactions. Two-factor interactions are confounded among themselves! This resolution is not so critical and very often used in the screening period.
- **Type V**: Main effects and two-factor interactions are not confounded among themselves. However, two-factor interactions are confounded with three-factor interactions. Usually completely uncritical and very often used also for optimization.

Higher resolutions (VI, VII, VIII, etc.) are possible (see Table 2.20). Further details regarding resolution are available in the literature [3].

2.3.4 Plackett-Burmann designs

As alternative to 2^{k-p}-designs also *Plackett-Burmann* designs can be used. In general these designs have a resolution of III. Thus, no confounding of main effects among themselves is present, but they are confounded with two-factor interactions. *Plackett-Burmann* designs are good suited for screening of a large number of influencing parameters and it is possible to create also designs for 12, 20, 24, 28, etc. factors. However, 2^{k-p}-designs offer only the possibility of designs for 4, 8, 16, 32, 64, 128, etc. factors and only linear models are possible. Thus, they are traditionally used in cases when only main effects are of interest. If the number of factor combinations is doubled, also plans with a resolution of type IV might be generated. For more details, please look into the literature [3].

2.3.5 DoE in colour measurement of a red-metallic base coat – 2^{6-1}-fractional factorial design

This example considers the influence of 5 different pigments and an effect-pigment mixture on the colour shade of a red-metallic base coat. Target value is the difference in lightness dL at the 45°-angle.

Due to the fact, that above 5 factors the number of experiments increases drastically, very often fractional factorial designs are applied. The number of experiments can be reduced, although the resolution of the design is still the same. Thus, no confounding of main effects among themselves is present and also no confounding with two-factor interactions.

Table 2.21: Factor settings for the example colour measurement of a red-metallic base coat

Factors	Low limit in [%]	High limit in [%]
White pigment (micr.)	0	2.0
White pigment	0	0.8
Red pigment 1	0	3.0
Pink	0	3.0
Red pigment 2	0	3.0
Effect pigment	0	3.0

Table 2.22: Design matrix for the example colour measurement of a red-metallic base coat: in case of a full factorial design 67 would be necessary. However, only those experiments were performed, where response values are available in the table

Number	Factors						Response
	White (micr.)	White	Red 1	Pink	Red 2	Effect	dL 45°
1	1	0.4	1.5	1.5	1.5	1.5	-4.59
2	0	0	0	0	0	0	1.25
3	2	0	0	0	0	0	-
4	0	0.8	0	0	0	0	-
5	2	0.8	0	0	0	0	2.09
6	0	0	3	0	0	0	-
7	2	0	3	0	0	0	-3.84
8	0	0.8	3	0	0	0	-3.84
9	2	0.8	3	0	0	0	-
10	0	0	0	3	0	0	-
11	2	0	0	3	0	0	-0.53
12	0	0.8	0	3	0	0	-0.64
13	2	0.8	0	3	0	0	-
14	0	0	3	3	0	0	-5.57
15	2	0	3	3	0	0	-
16	0	0.8	3	3	0	0	-
17	2	0.8	3	3	0	0	-4.62
18	0	0	0	0	3	0	-
19	2	0	0	0	3	1.5	-4.81
20	0	0.8	0	0	3	3	-3.79
21	2	0.8	0	0	3	3	-
22	0	0	3	0	3	3	-7.52
23	2	0	3	0	3	3	-
24	0	0.8	3	0	3	3	-
25	2	0.8	3	0	3	3	-6.94
26	0	0	0	3	3	3	-6.32
27	2	0	0	3	3	3	-

Table 2.22 (Continue)

Number	Factors						Response
	White (micr.)	White	Red 1	Pink	Red 2	Effect	dL 45°
28	0	0.8	0	3	3	3	-
29	2	0.8	0	3	3	3	-5.73
30	0	0	3	3	3	3	-
31	2	0	3	3	3	3	-8.94
32	0	0.8	3	3	3	3	-7.85
33	2	0.8	3	3	3	3	-
34	1	0.4	1.5	1.5	1.5	3	-4.81
35	0	0	0	0	0	3	-
36	2	0	0	0	0	3	1.22
37	0	0.8	0	0	0	3	2.32
38	2	0.8	0	0	0	3	-
39	0	0	3	0	0	3	-3.18
40	2	0	3	0	0	3	-
41	0	0.8	3	0	0	3	-
42	2	0.8	3	0	0	3	-2.54
43	0	0	0	3	0	3	-1.62
44	2	0	0	3	0	3	-
45	0	0.8	0	3	0	3	-
46	2	0.8	0	3	0	3	-0.67
47	0	0	3	3	0	3	-
48	2	0	3	3	0	3	-5.56
49	0	0.8	3	3	0	3	-4.7
50	2	0.8	3	3	0	3	-
51	0	0	0	0	3	3	-4.59
52	2	0	0	0	3	3	-
53	0	0.8	0	0	3	3	-
54	2	0.8	0	0	3	3	-3.7
55	0	0	3	0	3	3	-
56	2	0	3	0	3	3	-7.62
57	0	0.8	3	0	3	3	-6.71
58	2	0.8	3	0	3	3	-
59	0	0	0	3	3	3	-
60	2	0	0	3	3	3	-6.33
61	0	0.8	0	3	3	3	-5.44
62	2	0.8	0	3	3	3	-
63	0	0	3	3	3	3	-8.74
64	2	0	3	3	3	3	-
65	0	0.8	3	3	3	3	-
66	2	0.8	3	3	3	3	-7.92
67	1	0.4	1.5	1.5	1.5	1.5	-4.55

Properties of the experimental fractional factorial 2^{6-1}-design (see Table 2.22):

- Number of factors: 6 (see Table 2.21)
- Number of blocks: 1
- Number of target values: 1
- Number of experiments: 35, including 3 replicates in the centre point
- Randomization: yes

The pigments red 1 and red 2, pink and white show significant effects on the response dL 45°, but both red pigments are mainly important (see Figure 2.17 and 2.18). Significant is also the interaction between both red pigments. Hence, they are not independent from each other (Figure 2.19). The response surface indicates the sensitivity of the target value against variations of both pigment contents red 1 and red 2. Fluctuations in red 2 end up in much larger dL-differences than in case of red 1 (se Figure 2.19). Thus, DoE offers the possibility to study also aspects of robustness of recipes and to perform an analysis of sensitivity (see Figure 2.20 and Chapter 4.3). Beside this knowledge can also be used to make predictions in case of adjusting the colour shade.

This example demonstrates also the efficiency of fractional factorial designs. In case of a full factorial design with 6 factors and 3 replicates in the centre-point totally 67 experiments would be necessary. A 2^{6-1}-design with 3 replicates in the centre-point enables the same statements with only 35 experiments. A further reduction would be possible, but then the quality and validity of the predictions decreases. Such experimental designs cannot be analyzed with the straight forward methods of 2^2-design. In these cases so called response-surface methods based on regression analysis are used, which are described in Chapter 3.3. and 4.1.

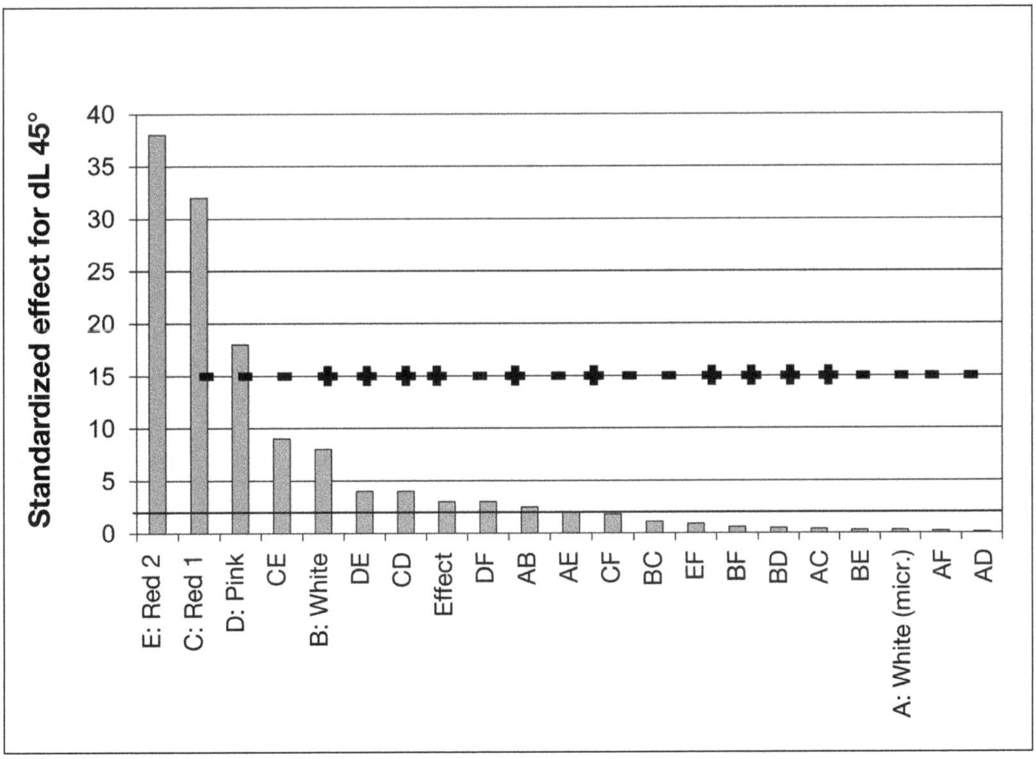

Figure 2.17: Pareto-diagram for the example colour measurement of a red-metallic base coat

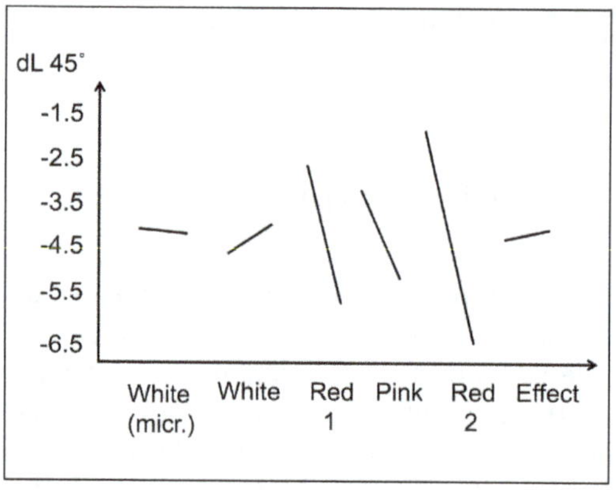

Figure 2.18: Main effect diagram for the example colour measurement of a red-metallic base coat

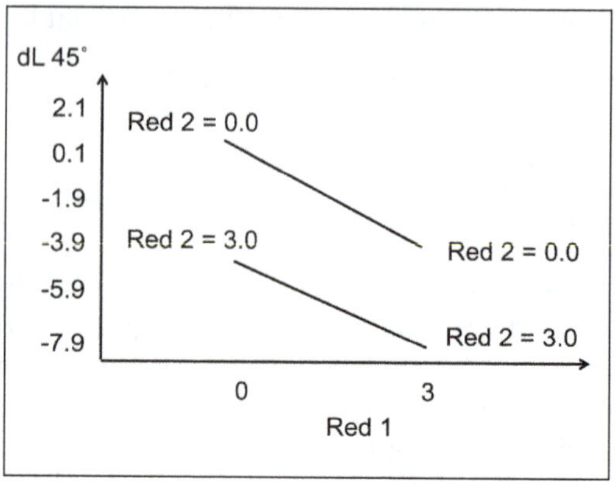

Figure 2.19: Interaction diagram for the example colour measurement of a red-metallic base coat

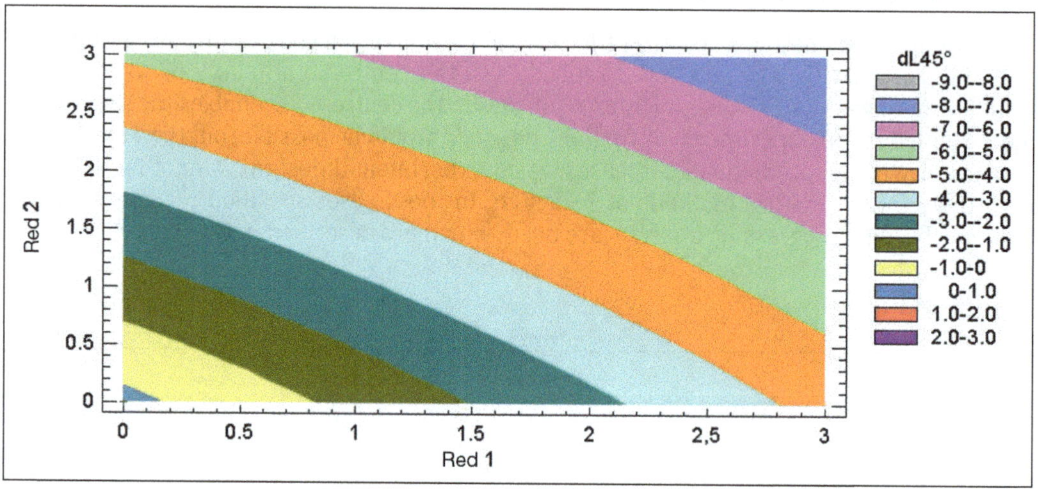

Figure 2.20: Response surface for the example colour measurement of a red-metallic base coat; white (micr.): 1 %, white: 0.4 %, pink: 1.5 %, effect: 1.5 %

2.4 Non-linear effect designs

If an influencing factor is only analyzed with very simple models based on two data points (see Figure 1.31a and b), there is a high chance to run the risk of overseeing non-linear behaviour. However, with an additional data point - which is an acceptable additional effort - it is possible, to check if all factors are linear or not (see also Chapter 2.2.4). If the difference is significant, at least one factor shows a non-linear behaviour. To get a sound knowledge in case of non-linear models at least three data points are necessary. Three strategies are possible:

- Central composite designs
- Three- and higher level designs like 3^k, or 3^{k-p}
- Mixed level designs
- *Box-Behnken* designs
- Optimal designs (D-optimal, etc.)

These different designs (very often called response-surface designs) are described in the next chapters.

tribution, but still typical methods of data analysis can be used. However, a very detailed description is available about the properties of each grade, which is always important to increase the degree of objectivity. If the evaluation is performed very subjectively, modelling is quite tough. Thus, ideally several persons evaluate the situation and an average value is used afterwards. In addition the scale-steps should not be too fine. It is also important that the scale of grades is regular and some marks should be available. Avoid just two levels like good and bad, because this excludes classical regression modelling! Thus, in principle priority have measurable responses, followed by a semi-quantitative multi-step classification. However, if only two qualitative steps "good" and "bad" are available, special methods like **logistic regression** can be applied. Data analysis is still poissible; further details can be found in literature [7].

It is also possible to deal with **qualitative factors** like "filter used – yes/no" or "produced in factory A or B", "batch 1 or 2". The design has to be adapted to the situation, but DoE can deal with such factors without any problems (see Chapter 2.3.2, Example blocking). However, data analysis by regression or modelling (response-surface methods) is not possible. If two-level designs are used, an analysis can take place as described in Chapter 2.2. If more levels are necessary, analysis of variance (ANOVA) (see Chapter 3.5) has to be applied. Thus, qualitative factors are not so critical than qualitative responses. At least it is a little trickier to interpret the coefficients compared to a situation with only quantitative factors or qualitative factors at only two levels.

▶ ***Basic principle:*** DoE enables to deal with qualitative influencing factors as well as responses. In case of qualitative target values it is necessary to get an objective evaluation by a large sample of replicates and an equal distance between the marks.

Example: Blocking of a clear coat as a function of the coalescent

The mixture of a hydrophilic and a hydrophobic coalescent as well as water should be optimized regarding blocking of a clear coat. The total amount of the three components should be max. 6 % (mixture design). Blocking is examined with a 4-stage classification (scale running from 0 = no blocking to 3 = complete sticking together, being 3 the worst).

Table 2.34: Experimental design regarding the example blocking

Name of experiment	Number of experiment	Amount of hydrophobic coalescent	Amount of hydrophilic coalescent	Amount of water	Blocking
1	4	0.33	0.33	0.33	0.7
2	9	0.33	0.67	0	1
3	1	0.67	0.33	0	0.7
4	11	0	0	1	0
5	2	1	0	0	0
6	12	0	1	0	2
7	3	0	0.67	0.33	1.2
8	7	0	0.33	0.67	0.7
9	10	0.33	0	0.67	0.5
10	6	0.67	0	0.33	0.5
11	5	0.33	0.33	0.33	0.7
12	8	0.33	0.33	0.33	0.7

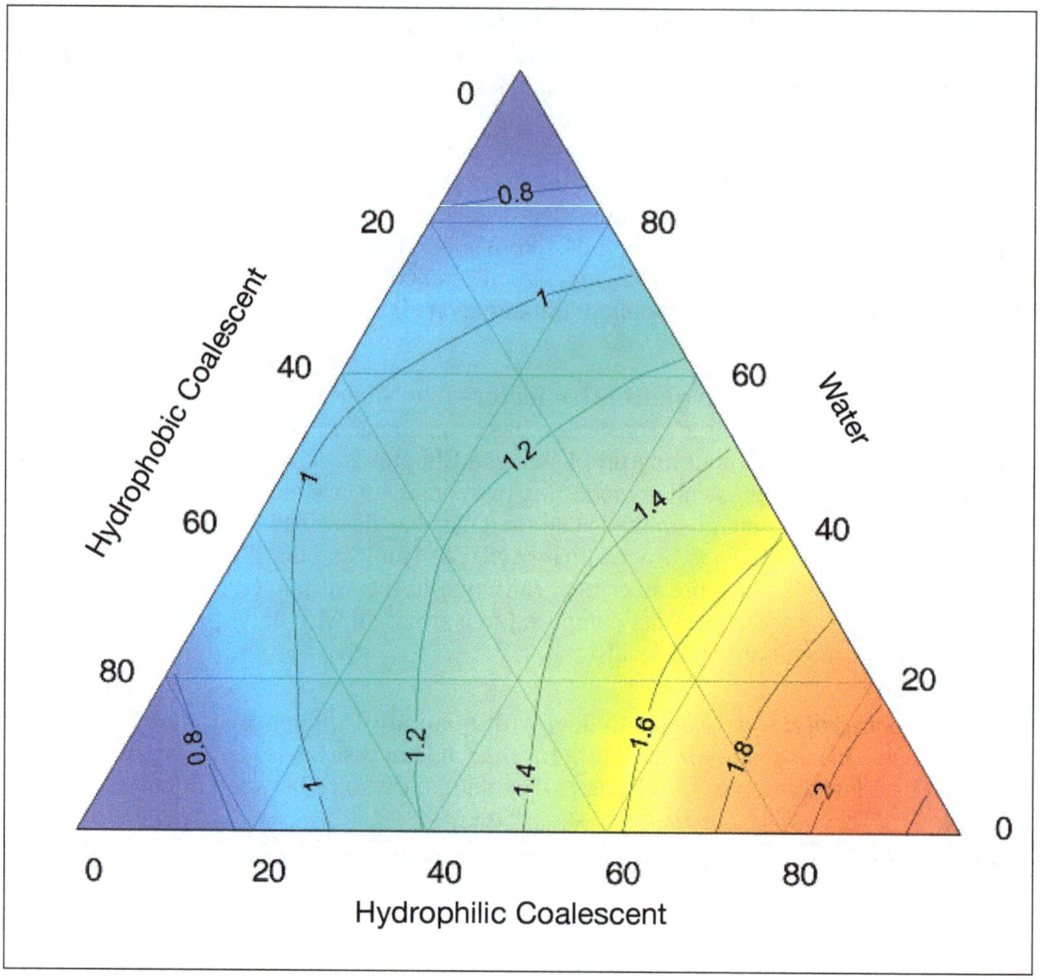

Figure 2.34: Blocking as a function of the influencing factors hydrophilic and a hydrophobic coalescent as well as water regarding the example blocking

Properties of the mixture design (see Table 2.35):

- Number of components: 3 (see Table 2.34)
- Number of target values: 1
- Number of experiments: 12
- Randomization: yes

The response surface shown in Figure 2.34 indicates that an increasing amount of hydrophilic coalescent decreases the blocking resistance. Good blocking is possible in two regions: on one hand the amount of both solvents is minimized for the benefit of water or on the other hand the content of hydrophobic coalescent is maximized. Such an analysis enables a perfect optimization of the recipe. In addition a sound knowledge about the robustness of the system is available. An introduction using trinary diagrams is given in Appendix 6.

Table 2.35: Factor settings regarding the example blocking

Factors	Low limit in [%]	High limit in [%]
Water	0	6
Hydrophilic coalescent	0	6
Hydrophobic coalescent	0	6

Example: Optimization of a spray automat regarding wetting of the edges

During the application of a conventional curing, waterborne pigment coating some unsatisfactory results occur in wetting the edges of a kitchen door. Only a high application amount of 200 g/m^2 enables an acceptable quality. However, this high amount of material ends up in an insufficient blocking resistance.

The control parameters of the spray automat are optimized in parallel regarding an application amount of 150 g/m^2, which offers a good blocking behaviour and a perfect wetting of the edges. In a preliminary step 9 parameters were evaluated by an FMEA (see Appendix 8). Also qualitative factors (see Table 2.36) are present. Target value is in each data point the minimal application amount for a perfect wetting of the edges. The determination is performed by variation of the conveyer belt velocity - an increase reduces directly the applied amount.

A full factorial design would be with 2^9 experiments impossible, even a fractional factorial 2^{9-2}-design needs 128 experiments. Also a further reduction with a simple fractional 2^{9-3}-design needs 64 experiments, but also a screening with resolution IV with a 2^{9-4}-design would need 32 experiments. Thus, a D-optimal design with 18 experiments was used and created by a software package. The power of this design is demonstrated impressively (see Chapter 2.4.5).

Properties of the D-optimal design (see Table 2.37):

- Number of components: 9 (see Table 2.36)
- Number of blocks: 1
- Number of target values: 1
- Number of experiments: 18
- Randomization: no

The most important effect on the application amount at optimal wetting of the edges has the factor working modus. Much better results can be achieved with the "pull-mode" than with the "pull/push-mode". Also gun-distance, gun-angle x-axis and material pressure are relevant. All three parameters should be adjusted at the lower level, because increasing ends up in higher application amounts. Other factors are not significant or have a very small effect. In the next step this design could be analyzed more in detail regarding these 4 parameters and their interactions. However, also at this stage an optimized operating area can be defined. A clear proof of the potential of DoE to reduce the experimental effort during screening with D-optimal designs.

Table 2.36: Factor settings regarding the optimization of a spray automat

Factors	Description	Unit	Low level -1	High level +1
A	Gun angle y-axis	[°]	25	35
B	Gun angle x-axis	[°]	35	60
C	Gun distance	[cm]	15	25
D	Spray nozzle approach angle	[°]	20	30
E	Working modulus	-	pull/push	push
F	Carriage speed	[m/s]	0.4	1
G	Spray nozzle size	[mm]	0.28	0.33
H	Material pressure	[bar]	100	150
I	Pressure atomized air	[bar]	1.5	2

Table 2.37: Design matrix regarding the optimization of a spray automat

Number	A	B	C	D	E	F	G	H	I	Minimal application amount in [g/m²] regarding a perfect wetting of the edges
1	-1	-1	-1	1	-1	1	-1	1	1	205
2	1	1	1	1	-1	1	1	-1	1	235
3	-1	1	1	1	1	-1	-1	-1	1	160
4	1	-1	1	-1	-1	-1	-1	1	1	240
5	1	-1	1	1	1	1	-1	1	-1	150
6	-1	-1	1	-1	-1	1	-1	-1	-1	225
7	1	-1	-1	-1	1	1	-1	-1	1	135
8	1	1	1	-1	1	-1	1	-1	-1	165
9	1	1	-1	-1	-1	1	-1	1	-1	215
10	-1	1	1	-1	1	1	1	1	1	180
11	-1	-1	-1	-1	1	-1	-1	1	-1	140
12	-1	1	-1	1	1	1	1	-1	-1	135
13	1	-1	-1	-1	-1	1	1	1	-1	212
14	-1	1	-1	-1	-1	-1	1	-1	1	225
15	1	1	-1	1	1	-1	1	1	1	155
16	-1	1	1	1	-1	-1	1	1	-1	240
17	-1	-1	1	1	1	-1	1	-1	1	146
18	1	-1	-1	1	-1	-1	-1	-1	-1	200

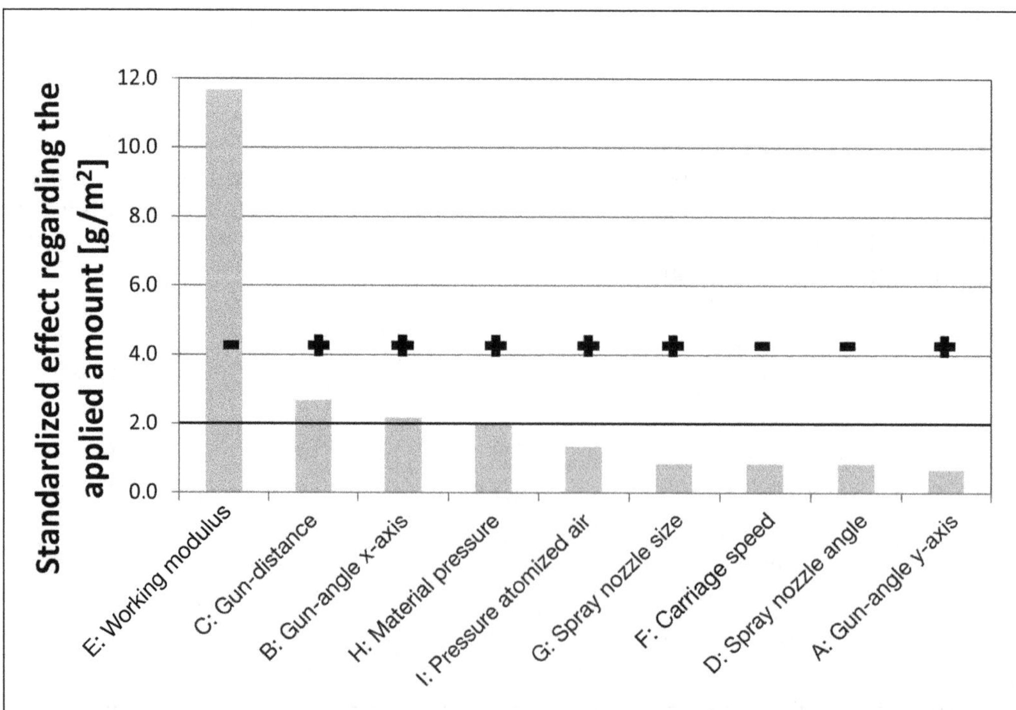

Figure 2.35: Pareto-Diagram for the experimental design regarding the optimization of a spray automat

2.7 *References*

General literature references can be found in the Appendix 9.

[1] K. Siebertz, D. van Bebber, T. Hochkirchen, Statistische Versuchsplanung, Springer, Heidelberg, 2010
[2] M. Adam, Statistische Versuchsplanung und Auswertung, Düsseldorf, http://mv.fh-duesseldorf.de, 2010
[3] W. Kleppmann, Taschenbuch Versuchsplanung - Produkte und Prozesse optimieren, 6th Edition, Hanser, 2009
[4] R. H. Myers, D.C. Montgomery, Response Surface Methodology, John Wiley, New York, 2th Edition, 2002
[5] C. Cornell, Experiments with Mixtures, John Wiley, New York, 3rd Edition, 2002
[6] R. D. Snee, Chemtech 1979, 702
[7] L. Braun, C. Morgenstern, M. Radeck, Prozessoptimierung mit statistischen Verfahren, Hanser Verlag, München, Wien, 2010
[8] M. Jäger, U. Schmidt, Farbe und Lack 2007, 113(8), 20
[9] P. R. Schalter, The Leader's Handbook, Mc Graw Hill, 1998

3 Number-crunching - nothing ventured, nothing gained in data analysis

After the experiments **data analysis and interpretation** start as next step (Table 1.7). Depending on the experimental design and on the problem, several different methods are applied itself or combined:

- Confidence interval in the *Pareto*-plot
- Regression analysis/response-surface-method (RSM)
- Analysis of variance (ANOVA)

However, beside mathematics common sense should be leading in this phase, because results have to be analyzed. The size and signs of effects must make sense and a plausible explanation for the effects should be present. In addition, data must give an answer to the original requests.

Data analysis involves in general three steps (Table 1.7):

- Primery evaluation of raw data
- Regression analysis and modelling
- Model interpretation and use of the model for further steps

3.1 Evaluation of raw data

3.1.1 Transformation

It has been emphasized already, many statistical methods assume that the data used are drawn from a normal population. Thus, testing for normality of distribution is important (by a simple histogram or a normal probability net (see Chapter 4, Appendix 3 or further literature [3, 4, 9]). If this behaviour is not the case, we are able to transform the data so that the data set is changed into a normal distribution. Significant deviations from the normal distribution would end up in completely irrelevant confidence intervals. Such transformations are regularly applied to the results of scientific experiments. Most common is plotting log x (instead of x), the **logarithmic transformation**. Details can be found in literature [3, 4, 9]. This transformation is very effective especially in case of right-skewed distributions with a longer tail on the right side and the mass of the distribution concentrated on the left side of the mean. It is important to re-transform the data after analysis (e.g. average) and be aware that the confidence interval is no longer symmetric. Transformation based on square root from data has not such a strong effect, but is still used for right-skewed distributions like the *Poisson's* distribution. In case of left-skewed data the quadratic transformation is suggested.

▶ ***Basic principle:*** Significant deviations from the normal distribution end up in irrelevant data analysis. Thus, transformation of original experimental data is applied to achieve a normal distribution.

Equation 3.10: $SQ_{Total} = (1.43 - 1.21)^2 + \ldots + (1.14 - 1.21)^2 = 0.13$

The average values of the groups are:

Avearge of all data at L+: 1.29
Avearge of all data at L-: 1.13

Thus, regarding the solvent the square sum of differences inside the group is:

Equation 3.11:
$SQ_{Rest} = (1.43 - 1.29)^2 + \ldots + (1.22 - 1.29)^2 + (1.1 - 1.13)^2 + \ldots + (1.14 - 1.13)^2 = 0.070$

The difference between the groups is (4 data points per group):

Equation 3.12: $SQ_{Effect, L} = 4 \cdot (1.29 - 1.21)^2 + 4 \cdot (1.13 - 1.21)^2 = 0.056$

Now we see that ANOVA provides us by decomposition of the total sum of squares SQ_{Total} at least with two estimates of the variance – one based on the inherent variability within the levels SQ_{Rest} and one based on the variability between the levels SQ_{Effect}. If there are no differences in the level means, these two estimates should be similar. If they are not, we can suspect that the observed difference must be caused by differences in the level means.

The amount of variance, which can be described by a certain effect is:

Equation 3.13: $$\frac{SQ_{Effect, L}}{SQ_{Total}} \cdot 100\,\% = \frac{0{,}056}{0{,}13} \cdot 100\,\% = 44.64\,\%$$

44.64 % of the variance can be explained by the effect of solvent L. What is about the remaining 55.36 %? Is this really just random error (noise)? No, because the system includes also other effects. Until now just the solvent has been evaluated. Also certain effects can be explained by the thickener and the surfactant (same calculation procedure). The total explainable effect is $SQ_{Effect} = SQ_{Effect, V} + SQ_{Effect, N} + SQ_{Effect, L}$). It is also possible to include interactions ($SQ_{Effect} = SQ_{Effect, V} + SQ_{Effect, N} + SQ_{Effect, L} + SQ_{WW(LV)} + SQ_{WW(LN)} + SQ_{WW(VN)} + SQ_{WW(LVN)}$), because they are also part of the system and can be explained. Summing up all square sums the total sum of squares SQ_{Total} can be calculated. If we calculate the ratio $\frac{SQ_{Effect}}{SQ_{Total}}$ for all other variables and interactions like for the solvent, Table 3.4 is generated.

Table 3.3: Experimental data of the oscillatory-rheological analysis of a waterborne pigmented coating

Number of experiment	Solvent L	Thickener V	Surfactant N	G`/G``
1	1	1	1	1.430
2	1	-1	-1	1.120
3	-1	1	1	1.100
4	-1	-1	-1	1.170
5	1	1	-1	1.400
6	-1	1	-1	1.090
7	1	-1	1	1.220
8	-1	-1	1	1.140

Comparing the data, sometimes higher values are realized than in other cases. Especially the two-factor interaction thickener/surfactant has a very small part in the total sum of squares. The result fits in a qualitative manner also to the *Pareto*-plot and the quantitative test result via the confidence interval is the same (see Chapter 1.6 and 3.2). However, are there any other effects significant? In addition, no quantitative probability estimate is available whether the value is just barely in the rejection region or whether it was very far in this region. The answer to this question is given by the ANOVA.

Table 3.4: Fragmentation of the variance to the parts which can be explained

	L	V	N	WW(LV)	WW(LN)	WW(VN)	WW(LVN)	G`/G``
	1	1	1	1	1	1	1	1.43
	1	-1	-1	-1	-1	1	1	1.12
	-1	1	1	-1	-1	1	-1	1.10
	-1	-1	-1	1	1	1	-1	1.17
	1	1	-1	1	-1	-1	-1	1.40
	-1	1	-1	-1	1	-1	1	1.09
	1	-1	1	-1	1	-1	-1	1.22
	-1	-1	1	1	-1	-1	1	1.14
Average of all data at + level	1.293	1.255	1.223	1.285	1.228	1.205	1.195	
Average of all data at – level	1.125	1.163	1.195	1.133	1.190	1.213	1.223	
SQ_{Effect}	0.056	0.017	0.0015	0.047	0.0028	0.00011	0.0015	
$\frac{SQ_{Effect}}{SQ_{Total}}$ **in %**	44.644	13.615	1.203	37.006	2.238	0.089	1.204	

Table 3.5: Formation of the sum of squares in case of two models[2]

	Model 1		**Model 2**	
	Degrees of freedom f	**SQ**	**Degrees of freedom f**	**SQ**
L	1	0.056	1	0.056
V	1	0.017	1	0.017
N	1	0.002	1	0.002
LV	1	0.047	1	0.047
LN	1	0.003	-	-
VN	1	0.000	-	-
LVN	1	0.002	-	-
Sum model	7	0.126	4	0.122
Error absolute	0	0	3	0.004
Error relative	-	0 %	-	3.2 %
Sum total	7	0.126	7	0.126

ANOVA enables to compare different models to describe the system. Fitting a full model with all parameters (all three main effects and the interactions: L, N, V, LV, VN, LN, LVN) in Table 3.4 will end up in the best description of the system (see model 1 in Table 3.5), because there is no space for errors (expressed by SQ_{Rest}). With 8 variables all coefficients of the model (intercept and 7 coefficients) are precise defined ($SQ = SQ_{Effekt,\,V} + SQ_{Effekt,\,N} + SQ_{Effekt,\,L} + SQ_{WW(LV)} + SQ_{WW(LN)} + SQ_{WW(VN)} + SQ_{WW(LVN)}$). However, if some replicates would be performed, a certain rest - model-independent - would remain due to random error (Figure 3.13).

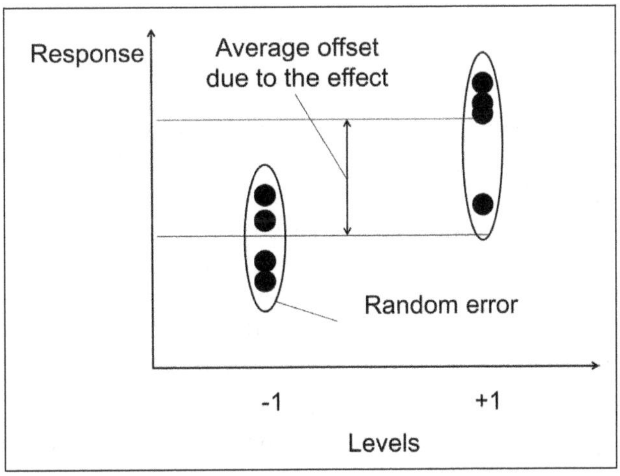

Figure 3.13: Splitting the variance into an explainable effect and the random error by noise

Table 3.6: Simple ANOVA for model 2 of Table 3.5 [2]

	f	SQ	F-factor	p-value
L	1	0.056	37.935	0.010
V	1	0.017	11.569	0.049
N	1	0.002	1.023	0.390
LV	1	0.047	31.445	0.013
Sum model. $SQ_{Regression}$	4	0.122		-
Error absolute $SQ_{Residuen}$	3	0.004		-
Sum total SQ_{Total}	7	0.126		-

Such an entire consideration is very often too complex. Models should describe reality with a sufficient accuracy. It's not a matter of true or false! Models are always false, because they offer only an approximation to reality (see also Chapter 1.7). Same is true in cartography. Depending on the application of a map this will be more or less reduced (see also Chapter 3.3).

If only three main factors and the dominant interaction between solvent and thickener is taken into account (reduced model 2 in Table 3.5), the open rest of variability is increasing (expressed by SQ_{Rest}). This is a basic principle: reducing the model will increase the error, independent of the noise. Thus, both terms are included in SQ_{Rest}. In parallel to the number of parameters also the so called degrees of freedom vary. If all parameters will be included, no degree of freedom is available. The neglected degrees of freedom f of the error is growing with a decreasing number of parameters (see Table 3.5). This has to be taken into account in the calculation procedure. Thus, the sum of squares is divided in a further step by the degree of freedoms (see Table 3.6, but f = 1 in this example). A factor which needs more degrees of freedom has to be respected in a different way, than one which needs only one degree of freedom.

Due to ANOVA the total sum of squares is partitioned into the effect, model or regression sum of squares and the error or residual sum of squares. Thus, it is possible to distinguish inside the total variance the amount of variance which can be explained by the model from the part of variance due to deviations from the model. Reducing the number of model parameters will increase $SQ_{Residuals}$. In case of single measurement without any replicates is:

Equation 3.14: $SQ_{Total} = SQ_{Effect} + SQ_{Rest} = SQ_{Regression} + SQ_{Residuals}$

For each parameter we calculate the ratio between SQ_{Effect} and the error (taking into account the degrees of freedom). In other words, we calculate the ratio of effect to noise. Increasing ratio indicates a growing significance of the effect. But which effects are significant?

A statistical test is necessary to answer this question (see Appendix 5). In principle we compare two sums of squares, which can be compared with the standard deviation (see Appendix 2). Thus, we test the difference between two variances. The significance tests so far described in this book are used for comparing means (t-test, see also Appendix 5). In this case it is important to compare variances (standard deviation square), which is usually done by the so-called **F-test** (developed by *R. A. Fisher*). This test distinguishes between equal variances and too great differences of SQ_{Effect} and $SQ_{Residuals}$, which cannot be explained by random variation.

A very common and useful criterion regarding the significance is the so-called **p-value** (see also Appendix 5). DoE- and statistic-software packages have widely adopted this value. Please look for further details into Appendix 5, in this chapter it is only important how to deal with this value in reality:

- p-value < 0.001: very strong significant effect
- 0.001 < p-value < 0.01: strong significant effect
- 0.01 < p-value < 0.05: significant effect
- p-value > 0.05: not significant effect

Table 3.6 is the result applying this procedure to the example. The conclusion from this analysis is that the effect of solvent and the interaction of solvent and thickener is significant (Table 3.6). A p-value equal to 0.049 indicates an effect of the thickener slightly below the limit of significance with 0.05. If in reality the solvent has no effect, the probability to get still a ratio 37.9 (= f-factor) of the two sum of squares $SQ_{Effect, L}$ and $SQ_{Residuals}$ is 1 % (α-error, see Appendix 5). In case of the thickener the probability is 4.9 % and the surfactant has 39 % probability.

As already emphasized, $SQ_{Residuals}$ has to be partitioned into:

- The variability due to random error (noise) and
- The variability due to an incorrect model, because some parameters are not included

Table 3.7: Evaluation of the regression model, which does not take all effects and interactions into account (see right side of the table) [2]

	Model 1		Model 2	
	f	SQ	f	SQ
L	1	0.056	1	0.056
V	1	0.017	1	0.017
N	1	0.002	1	0.002
LV	1	0.047	1	0.047
LN	1	0.003		-
VN	1	0.000		-
LVN	1	0.002		-
$SQ_{Residuals}$	0	0.000	3	0.004
$SQ_{Regression}$		0.126		0.122
SQ_{Total}		0.126		0.126
R^2		1		0.970

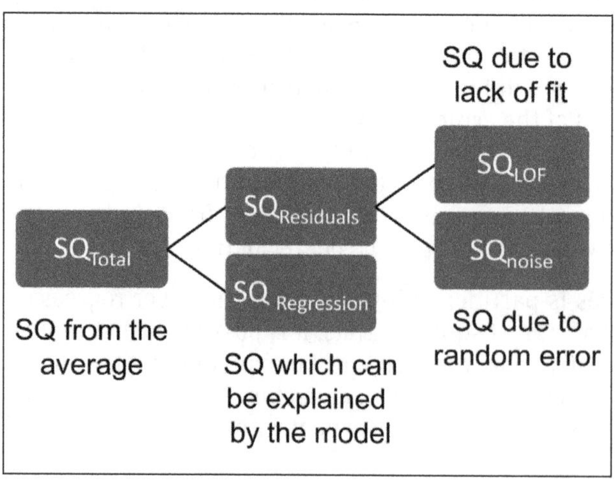

Figure 3.14: Splitting up the whole variance into several parts

As long as replicates are available, it is possible to distinguish between these two parts and the variance due to a low goodness of fit (so called **lack of fit, LOF**) can be calculated. The remaining rest is the so called noise (= **pure error, PE**) (see Figure 3.14).

Equation 3.15: $SQ_{Total} = SQ_{Regression} + SQ_{Residuals} = SQ_{Regression} + SQ_{LOF} + SQ_{Noise}$

Thus, finally ANOVA is able to evaluate also the regression model, because $\frac{SQ_{Regression}}{SQ_{Total}}$ is equal to the coefficient of determination R^2 (commonly also Goodness of fit, see Chapter 3.3 and Table 3.7). All SQ-values in Table 3.7 can be compared by an F-test: $SQ_{Regression}$ with $SQ_{Residuals}$ or SQ_{LOF} with SQ_{Noise}, etc. A correct model is indicated by a very small p-value for regression and a very large p-value for LOF.

This ANOVA has been calculated with three main components and the dominant interaction between solvent and thickener. The effect of the surfactant is with a p-value already insignificant. Thus, a certain model was the base for all calculations. With another model with more or less factors all data would be different. Also a non-linear model would be possible (solvent^2). The ANOVA-method is completely flexible and is recommended to check several models regarding residuals, lack of fit and goodness of fit. Not only significant parameters from the *Pareto*-plot end up in a correct model.

▶ ***Basic principle:*** Analysis of variance (frequently abbreviated to ANOVA) is an extremely powerful tool to separate any variation in the response which is caused by altering one factor, from variations due to random error (significant difference between two values). In case of models ANOVA distinguish between the amounts of variance which can be explained by the model from the variance due to deviations from the model. In addition, the deviation will be partitioned into the variability due to random error (noise) and the variability due to an incorrect model, because some parameters are not included (so called lack of fit). Reducing the model will increase the error, independent of the noise.

3.5.2 *Example: Colour measurement of a base coat – ANOVA*

In an experiment the influence of 4 different pigments on the colour shade of a blue-metallic base coat is studied. Two different models are compared.

Properties of the screening design (see Table 3.9):

- Number of components: 4 (see Table 3.8)
- Number of blocks: 1
- Number of target values: 1
- Number of experiments: 19

The response db 45° is mostly affected by the black and violet pigment. The influence of pink is marginal and no interaction is significant regarding the *Pareto*-diagram (see Figure 3.15). The same result can be achieved by the ANOVA-table (see Table 3.10). A p-value of 0.035 indicates only a weak significance for pink.

Table 3.8: Parameters incl. experimental area regarding the example colour measurement of a blue-metallic base coat

Factors	Low level in [%]	High level in [%]
Black pigment	0.0	0.44
White pigment	0.0	0.88
Pink pigment	0.0	1.36
Purple pigment	0.0	0.52

Model 1 fits very well the data set, because the p-value for the lack of fit is with 0.15 quite high (see Table 3.10). The alternative reduced model with less param-

Table 3.9: Experimental design regarding the example colour measurement of a blue-metallic base coat

Number	Factors				Response
z	Black	White	Pink	Violet	db 45°
1	0.22	0.44	0.68	0.26	1.14
2	0	0	0	0	-0.65
3	0.44	0	0	0	3.02
4	0	0.88	0	0	-0.44
5	0.44	0.88	0	0	2.96
6	0	0	1.36	0	-0.89
7	0.44	0	1.36	0	2.86
8	0	0.88	1.36	0	-1.01
9	0.44	0.88	1.36	0	2.78
10	0.22	0.44	0.68	0.26	1.37
11	0	0	0	0.52	-1.08
12	0.44	0	0	0.52	2.74
13	0	0.88	0	0.52	-1.13
14	0.44	0.88	0	0.52	2.62
15	0	0	1.36	0.52	-1.18
16	0.44	0	1.36	0.52	2.48
17	0	0.88	1.36	0.52	-1.28
18	0.44	0.88	1.36	0.52	2.48
19	0.22	0.44	0.68	0.26	1.11

eter can be evaluated in the same way, because only insignificant parameter has been eliminated (see Table 3.11). The p-value of the lack of fit is even higher (0.295), but the goodness of fit is equal.

This example emphasizes how ANOVA can be used to compare different models and fit them to the requirements. In addition, it is shown, how a colour shade is affected by varying parameters (see Figure 3.16). Only long-time employees have such knowledge available without any systematic analysis. However, this knowledge, with is more like experience, is mostly not quantitative always subjective. Thus, mixing colours is commonly time consuming and cost intensive. With the possibility for predictions based on models, internal processes can be optimized and technicians get helpful information for their work.

Table 3.10: Analysis of variance of db 45°-model 1 regarding the example colour measurement of a blue-metallic base coat

Factors	Sum of squares	f	Calculation including the degrees of freedom	F-ratio	p-value
A:Black	57.228	1	57.228	4061.92	0
B:White	0.0034	1	0.0034	0.24	0.6563
C:Pink	0.189	1	0.189	13.46	0.0350
D:Violet	0.542	1	0.542	38.47	0.0084
AB	0.0055	1	0.0055	0.39	0.5769
AC	0.0034	1	0.0034	0.24	0.6563
AD	0.0054	1	0.0054	0.39	0.5785
BC	0.0089	1	0.0089	0.63	0.4844
BD	0.0063	1	0.0063	0.44	0.5526
CD	0.011	1	0.011	0.77	0.4443
Lack-of-fit	0.316	6	0.053	3.74	0.1531
Random error	0.042	3	0.014		
Total (Corr.)	58.503	19			

R^2 = 99.3872 %

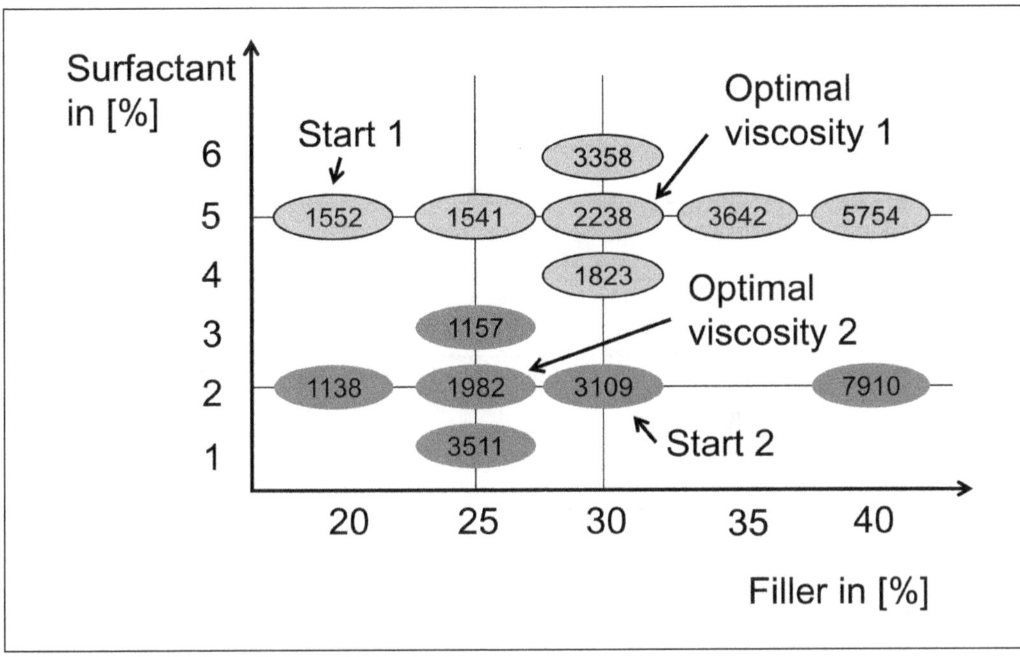

Figure 1.11: The result of an optimization depends in case of a non-linear system on the starting point. This figure shows the example of optimizing the viscosity of the filler concentrate (numbers in the data points show the viscosity in mPas). Two different starting points ends up in two different optimized recipes.

1.4.2.2 Non-linear effects and domain dependence

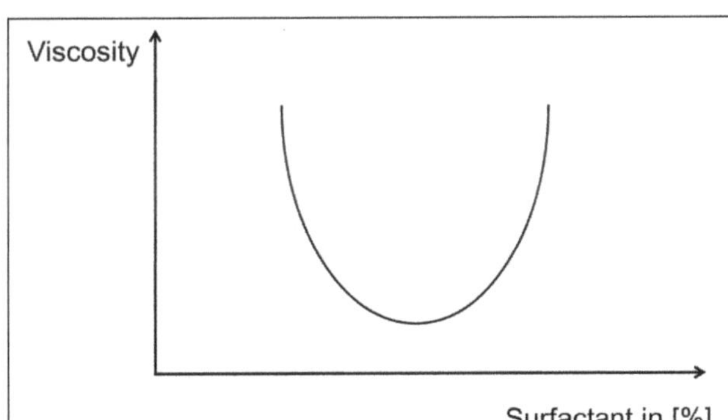

Figure 1.12: Viscosity versus surfactant concentration for a waterborne pigment-slurry

In Chapter 1.4.1 the experiments started with a composition of 5 % surfactant and 20 % filler and end up - regarding the target value viscosity - in an optimized composition of 30 % filler and 5 % surfactant. If another starting composition will be used, like 30 % filler and 2 % surfactant, usually is assumed that the experiments will end up at the same optimum. However, this is not the case and an optimum of 26 % filler and 2 % surfactant is resulting (see Figure 1.11). Therefore, the result is depending on the starting point! In principle this is positive, because 4 % less solid content can reduce the cost of raw materials and can improve also transparency. But what is the reason for this behaviour?

Everybody love to work with linear effects and tend to see the world also in this manner. As often as possible, we use a straight line to describe the system behaviour. For sure, this is comfortable when two factors are always in the same ration to each other [4]:

- The mass of a liquid is proportional to its volume, the factor is the density.
- The light switch is on and the light is on, the light switch is off and the light is off.

Unfortunately real life is more complex and deviations from the linear behaviour are very often the case. You can switch on the light, but it will be still dark, if the electric cable is

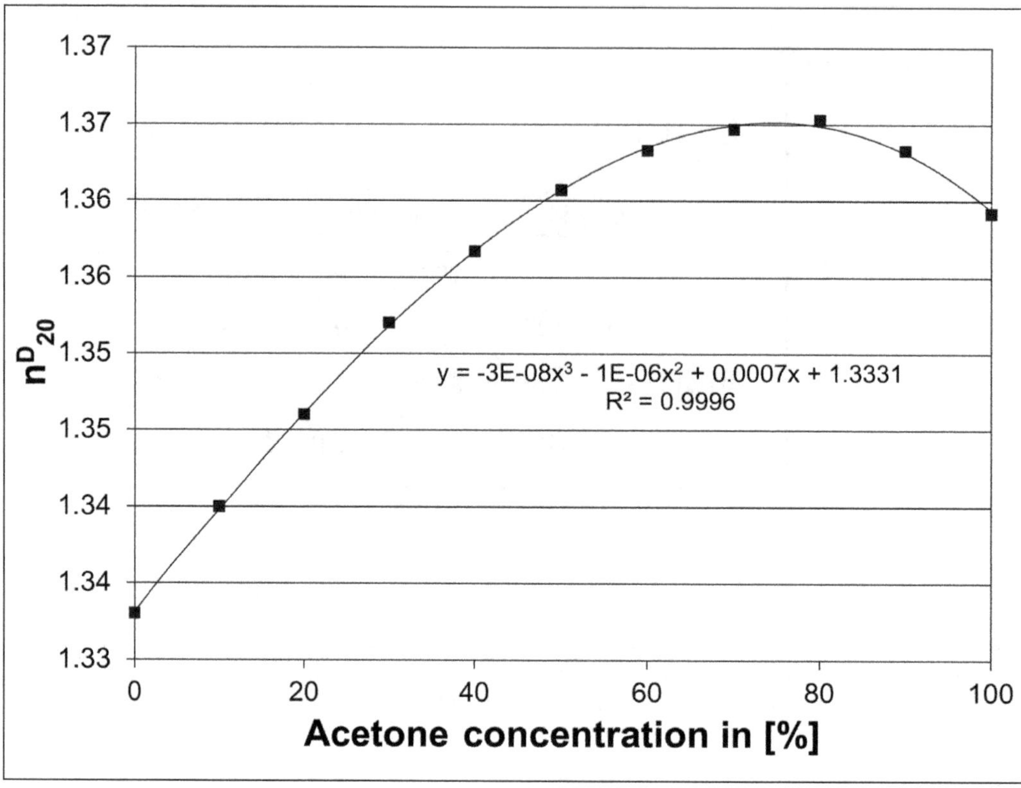

Figure 1.13: Refractive index of a acetone/water-mixture vs. concentration of acetone in wt.-%

not fixed, the electricity bill is not paid, or a blackout is present [4]. In case of **non-linear systems** the effect is not directly proportional in every region to the input signal.

You have never seen such behaviour in coating science? A few examples will be described in this book. However, already the example of the filler concentrate supposes such behaviour. This is not surprising and also from a perspective of coating formulation comprehensible. It is well known, that the optimal surfactant concentration in waterborne pigment slurries can be found very easy by using the fact, that the viscosity of the slurry has a minimum, when the pigment surface is completely occupied with additive (see Figure 1.12). For certain filler contents and certain viscosity target values, the optimal recipe will depend on the starting point (see Figure 1.11 and 1.12).

Today it is already assumed, that non-linear effects exceed classical predictable linear properties many times [5]. Just think about the following examples:

- Viscosity versus time after mixing the isocyanate hardener with the basic waterborne coating component - during pot life viscosity passes very often a minimum.
- Particle size or gloss improves during dispersing time due to fragmenting the pigment agglomerates. In contrast to that the contrast ratio passes in case of titanium dioxide a minimum, because there is no correlation of this value with the decrease of the agglomerates. The contrast ratio depends on the particle size distribution in the visible region. Thus, the properties of the coating system change non-linear [6].
- Tinting strength of a pigment is increasing with increasing amount of surfactant and is reaching a saturation value after a certain concentration.
- The refractive index in a mixture is a non-linear function of the mixing ratio (see Figure 1.13).

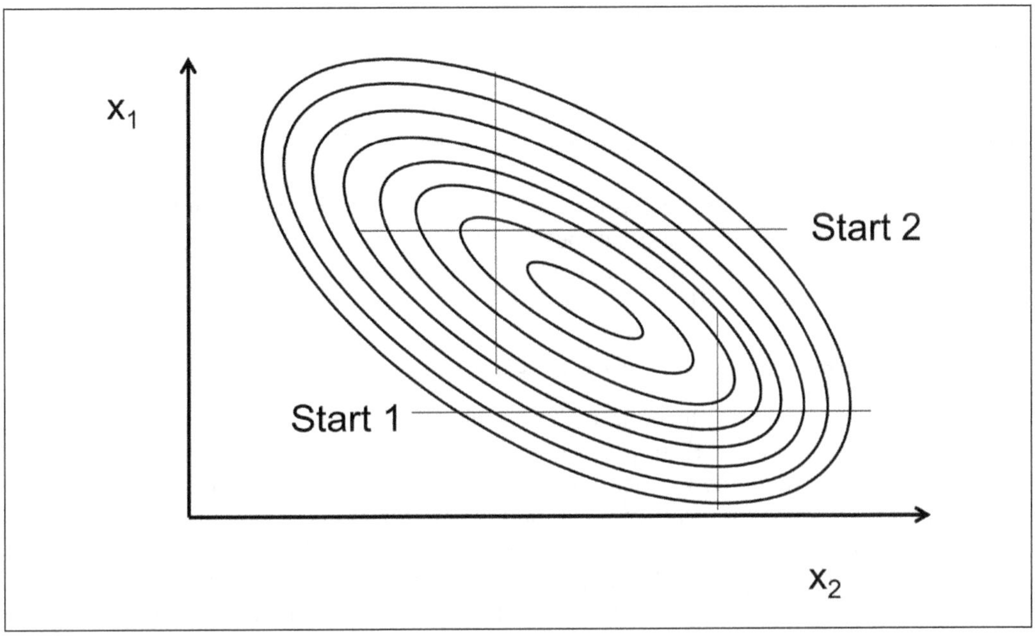

Figure 1.14: Searching for the optima of a system starting at two different starting points by the one-factor-at-a-time method (the response is shown by contour lines)

Thus, non-linear behaviour is very often present in coatings technology and it is very important to work with methods, which can deal with this issue. Design of experiments offers some possibilities for that and illustrates the presence of such behaviour (see Chapter 2.4).

The one-factor-at-a-time method and the intuitive experimentation have some limits in case of non-linear effects, because:

- Due to the variation of just one factor it is impossible to realize important correlations in the system. Each variation refers to one starting point. It is ambiguous, how the system reacts, when another starting point is chosen. For the interpretation of the results it has to be assumed that the effect of one factor is independent of the others. However, reality is often different, because non-linear effects are present and the optimal target value depends on the starting point (see Figure 1.11)!
- Probably optimal target points will not be identified. Even with many experiments, the one-factor-at-a-time method offers just a local analysis of the process behaviour (see Figure 1.14).
- The information is probably not robust against disturbance factors due to the local focus of the consideration.
- Interactions (see Chapter 1.5.2) cannot be identified in the system.

▶ ***Basic principle:*** Non-linear behaviour is very often present in science and technology. In case of non-linear systems, the result depends on the starting point of the experimental series. The classical approach performing experiments has some limits in case of non-linear systems and show a domain dependence.

1.4.2.3 Universality of the statements

If only one factor is varied at a time (independent if randomly or systematically), very detailed information can be generated about the single factor. However, only an incomplete description of the experimental area is achieved. This fact has already been described in

Chapter 1.4.2.2 about non-linear effects and in Figure 1.11. From each experimental series - independent which one - it cannot be derived, that the response depends on the starting point. It cannot be observed by performing series 1, that there is another formulation, which is much better regarding transparency and raw material costs (series 2). Design of experiments offers considerable better possibilities (see Chapter 1.5.2 and 1.7).

▶ ***Basic principle:*** The classical approach offers very detailed statements about effects of single factors. However, it is not possible to generate a general statement about the effects in the whole experimental area. Knowledge about valuable correlations in the system (e.g. non-linear effects) will be hidden.

1.4.2.4 You desire, we play - multiple responses

The big challenge in coatings formulation is searching for an optimized and balanced relation of all coating properties. High-performance products have always a certain advantage in some aspects. However, by implication it is assumed, that all other properties fulfil the state of the art. Therefore, in coatings formulation rarely just one target (e.g. scratch resistance) has to be optimized. Pot life, chemical resistance, ability for sanding, elasticity, raw material cost, etc. must also be fulfilled **(= multiple response optimization)**. Think about people working in the coloristic department during designing a metallic automotive OEM coating. A lot of experience is necessary to fulfil all 15 coloristic targets (da, db, dL for the angles 15°, 25°, 45°, 75° and 110°) with 4 or even more effect pigments and colour pigments. Design of experiments offers an approach for such situations (see Chapter 4.2). The classical approach is limited very soon in this case, because correlation of high complexity has to be realized by cross-linking completely different aspects. Very often also conflicting targets are present. Adjustments which are useful for one property might have negative effects on other properties.

▶ ***Basic principle:*** The classical approach cannot be used for multiple response optimizations in case of complex correlations and/or many influence factors.

1.4.2.5 The gain of knowledge and new information is too slow

Time is money and time is always short in research laboratories. Thus, it is in the interest of all people, that experiments are performed as time-saving as possible and to fulfil the economic principle: The volume of information has to be maximized regarding the effort of experiments. If the classical approach is used, information is growing linear. Each experiment gives us a new piece of the puzzle and a global knowledge is available only after almost all experiments.

▶ ***Basic principle:*** The gain of information is linear in case of classical methods. This is to slow for modern R&D-strategies and the demands of today's economics.

The aim is to use methods, which generate as much knowledge as possible (see Figure 1.15). The ***Pareto*-principle** (*Vilfredo Frederico Pareto,* Italian engineer, economist and sociologist, 1848 to 1923), or **80 to 20 rule,** states that, for many events, roughly 80 % of the effects on the response are caused by 20 % of the investigated factors, because these are the most important ones. So, with 25 factors approximately 5 have a real effect. As an example, everybody knows that a prototype with 80 % of the properties can be developed very soon in 20 % of the project time. However, the remaining 20 % of the results cause most of the work (e.g. by optimization) and consumes 80 % of the project time [7]. The ***Pareto***-principle can also be applied for several situations in the daily life [7]:

- We use in 80 % of our time just 20 % of our clothes. 80 % of our clothes are used only in 20 % of our time. Think about your favourite sweater, etc.
- 80 % of time inside our flats or apartments we move only on 20 % of the total available flooring area. We go always at the same way into the kitchen, read the newspaper at the same place, etc.

Thus, an additional basic principle for design of experiments can be derived:

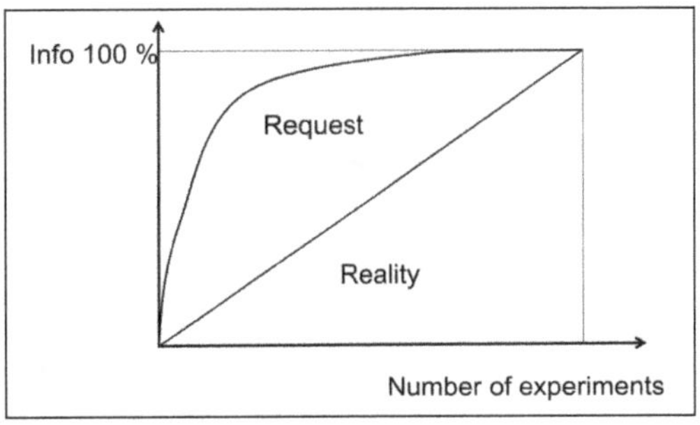

Figure 1.15: Volume of new information vs. number of experiments

▶ ***Basic principle:*** The aim in design of experiments is to complete 80 % of a certain task with 20 % of the total number of experiments.

1.5 Design of experiments - what's that?

1.5.1 Design, factors and effects

The planning and conduction of statistical analysis (also termed "design of experiments" or, abbreviated, DoE) commence at the previously mentioned stages, and therefore gives access to a huge economic potential. But at this stage one question is obvious: what is design of experiments in detail? We consider once more the example of developing a filler concentrate. In contrast to Chapter 1.4.1 experiments are performed only at two concentration levels (meaning low and high amount of surfactant like 2 and 4 % or 20 and 40 % filler).

In case of the traditional one-factor-at-a-time method the design will look like in Figure 1.16a. In case of DoE the set of experiments will analyze all possible combinations of both factors surfactant and filler according to a certain pattern. Usually the experiments are laid out in a symmetrical fashion. But how many experiments will be necessary? Four experiments, as shown in Table 1.3. Either both factors are at high or

Table 1.3: Design matrix with possible combinations in case of the experiments with two factors and two levels

Number of experiments	General identification	Factor A	Factor B
1	(1)	-	-
2	A	+	-
3	B	-	+
4	ab	+	+

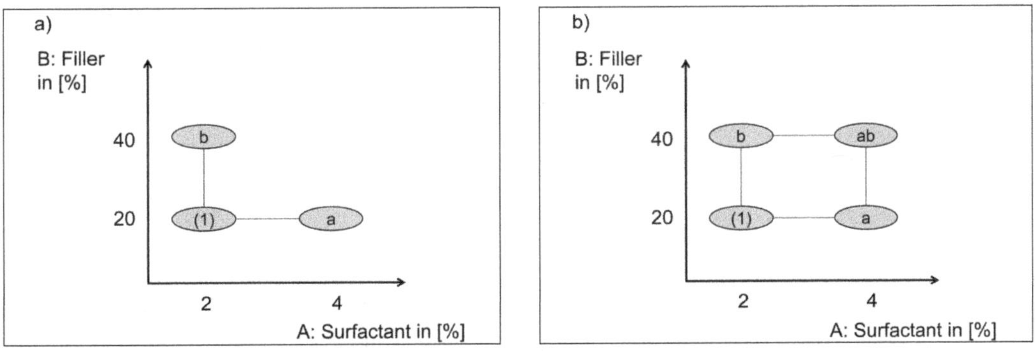

Figure 1.16: Geometric presentation of the experimental design for the filler concentrate: a) one-factor-at-a-time method, b) design of experiments with a two-factor factorial design

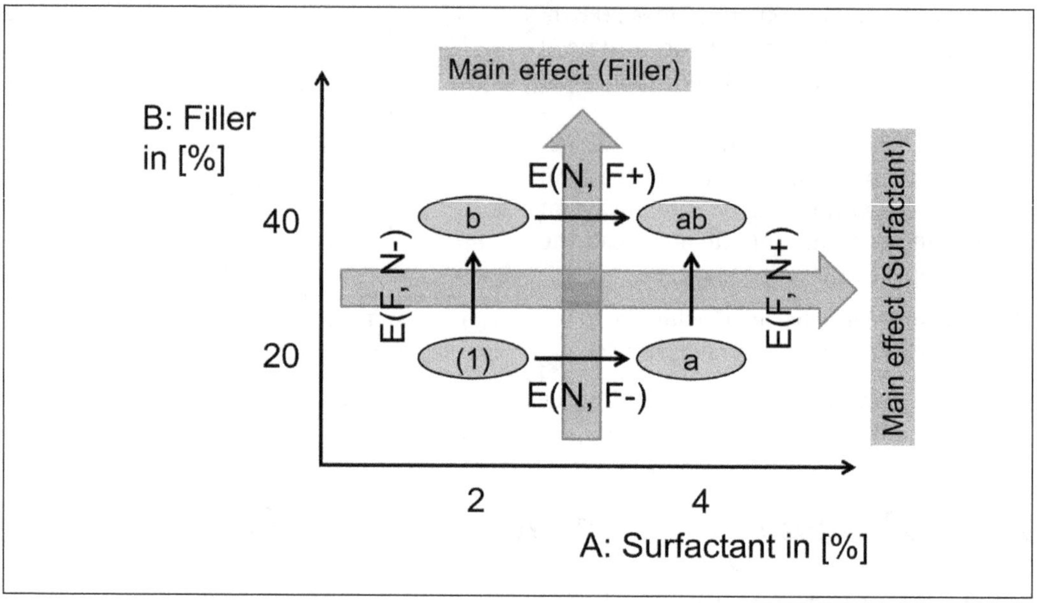

Figure 1.17: Development of a filler concentrate: Presentation of the analysis with effects and main effects

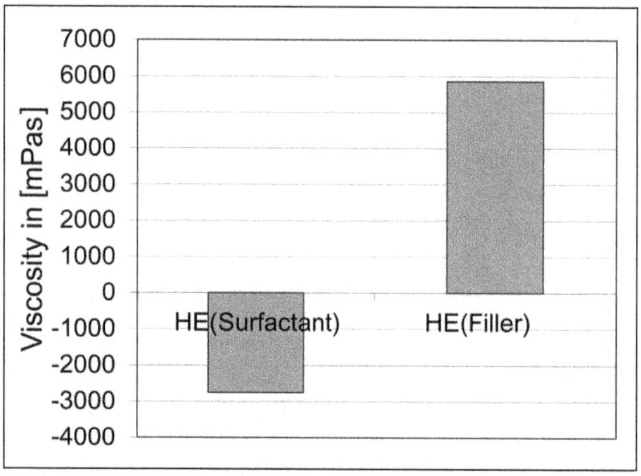

Figure 1.18: Main effect diagram in case of the example filler concentrate

low level (low-low, high-high), or one of the factors is high and the other low (low-high, high-low). So, there is an additional fourth experiment in contrast to the one-factor-at-a-time method in which both factors are varied simultaneously. This design is called a **factorial design.** However, if we look more in detail, this additional experiment offers a lot of advantages, which will be described in the following analysis.

If both factors are fixed at two levels and the response viscosity is measured, a design as shown in Figure 1.16b and Table 1.4 can be concluded and geometrically interpreted as a square. Hence the experimental region is said to be of regular geometry. It is in general strongly recommended to make graphical presentations of the design, because they are much more instructive as listing all experiments in tables. The important point is that each data point has its own identification and corresponds to one experiment, which will be used in the analysis step.

Table 1.4: Factorial design for two factors at two levels

Number of experiments	General identification	Factor		Response
		Surfactant (A)	Filler (B)	
1	(1)	-	-	2360 mPas
2	a	+	-	900 mPas
3	b	-	+	9500 mPas
4	ab	+	+	5470 mPas
Average value (= a_0)				4558 mPas

The measured data of the response viscosity from Table 1.4 will be analyzed as shown in Figure 1.17. The first step is the calculation of the **effects E for each factor at each level.** They match directly with the difference of the response values.

Effect E from the surfactant (N) at a low level of the filler (F-):
Equation 1.2: $E(N,F-) = a - (1) = 900 \text{ mPas} - 2360 \text{ mPas} = -1460 \text{ mPas}$

Effect of the surfactant at a high level of the filler:
Equation 1.3: $E(N,F+) = ab - b = 5470 \text{ mPas} - 9500 \text{ mPas} = -4030 \text{ mPas}$

Effect of the filler at a low level of the surfactant:
Equation 1.4: $E(F,N-) = b - (1) = 9500 \text{ mPas} - 2360 \text{ mPas} = 7140 \text{ mPas}$

Effect of the filler at a high level of the surfactant:
Equation 1.5: $E(F,N+) = ab - a = 5470 \text{ mPas} - 900 \text{ mPas} = 4570 \text{ mPas}$

The **main effect HE** is the average increase of a response due to varying one factor from its low level to its high level. It can be calculated from the average difference in response values when moving from low to high level and keeping the other factors at a constant level:

Equation 1.6:
$$HE(\text{Filler}) = \frac{E(F,N-) + E(F,N+)}{2} = \frac{(b - (1)) + (ab - a)}{2} = \frac{7140 + 4570}{2} \text{ mPas} = 5855 \text{ mPas}$$

Equation 1.7:
$$HE(\text{Surfactant}) = \frac{E(N,F-) + E(N,F+)}{2} = \frac{(a - 1) + (ab - b)}{2} = \frac{-1460 + (-4030)}{2} \text{ mPas} = -2745 \text{ mPas}$$

On average, an increase of the surfactant concentration leads to a reduction of the viscosity and an increase of the filler content leads to greater thickening. From a perspective of coating formulation, this is comprehensible.

The effects can be illustrated in a so called main effect diagram (see Figure 1.18). Software packages use very often another way, which is shown in Figure 2.16. Real effects are given by the coefficients of the regression model. Details are described in Chapter 1.7 and 3.3.

In alternative to the analysis used until now, main effects can be calculated also from the average of the response at all data points at the higher and lower level and a subtraction of these two average values afterwards (see Equation 1.8). This way of calculation is especially helpful when dealing with more than two factors (see Chapter 2.2.2).

Equation 1.8:
$$HE(\text{Surfactant}) = \frac{E(N,F-) + E(N,F+)}{2} = \frac{(a - 1) + (ab - b)}{2} = \frac{ab - b + a - (1)}{2}$$
$$= \frac{ab + a}{2} - \frac{b + (1)}{2} = \frac{5470 + 900}{2} - \frac{9500 + 2360}{2} = -2745 \text{ mPas}$$

▶ ***Basic principle:*** The main effect HE is the average increase of a response due to varying one factor from its low level to its high level. It can be calculated from the average difference in response values when moving from low to high level and keeping the other factors at a constant level. The effects can be illustrated in a so called main effect diagram.

In contrast to the one-factor-at-a-time design a much more general statement about the experimental area (and the system) is possible by DoE, because the factors are analyzed at all levels. The already described alternative factorial design of experimental approach is balanced and objective, no combination has more importance than another. The one-factor-at-a-time method prefers all experiments with a low concentration of the surfactant, because two data points are analyzed at this level (Figure 1.16a). In case of linear behaviour and a good reproducibility of the measurements, with this approach much more deep going statements can be achieved. However, the main effects calculated by the DoE approach offer a general explanatory power and therefore much more possibilities.

▶ ***Basic principle:*** In contrary to the classical approach, design of experiments offers in the whole experimental area a general, objective and balanced explanatory power. DoE deals with main effects of each influencing parameter, which describe the average increase of a response due to variation of the influencing factor inside the experimental area.

In case of this introductive example just two factors has been used. In the following chapters also examples with much more factors are analyzed (starting from Chapter 2.2.2 and 2.3).

1.5.2 Interactions

Already *Aristotle* said, that "the whole is greater than its parts". Looking on the effects, it is obvious, that the extension of each effect depends highly on the level of the factor. When this occurs, there is an **interaction between the factors.**

- Effect of the surfactant at a low level of the filler: - 1460 mPas
- Effect of the surfactant at a high level of the filler: - 4030 mPas

Especially in chemistry interactions have a significant importance in case of non-linear effects (see Chapter 1.4.2.2), but are very often ignored. Think about the synergy between UV-absorber and radical scavenger (e.g. HALS-additives). The whole is also in this case more than only the single effects.

From a perspective of coating formulation, also this effect is comprehensible. The surfactant will wet much more surface of the filler at a low level of the filler concentration. Thus, the effect on viscosity will be different from that one at a high level. However, interactions are not correlations, because the filler content is not changing due to the variation of the surfactant concentration. Only the influence of the surfactant on the response (e.g. viscosity) depends on the filler content, the adjustment of the second factor.

▶ ***Basic principle:*** Essential statements regarding interactions can only be generated with the DoE approach due to additional data point at high filler and surfactant concentration and the variation of both factors simultaneously. In addition, the one-factor-at-a-time method or the intuitive method offers not such a deep inside into the system.

It is state of the art to calculate a **classification number for the interaction WW.** In contrast to the main effect it's the difference from the single effects like shown in Equation 1.9. The interaction WW(FN) is equal to WW(NF) (see Equation 1.9).

Equation 1.9:

$$WW(FN) = \frac{E(N,F+) - E(N,F-)}{2} = \frac{-4030 - (-1460)}{2}\,\text{mPas} = \frac{(ab - b) - (a - (1))}{2}$$

$$= \frac{(ab - a) - (b - (1))}{2} = \frac{E(N+,F) - E(N-,F)}{2} = WW(NF) = -1285\,\text{mPas}$$

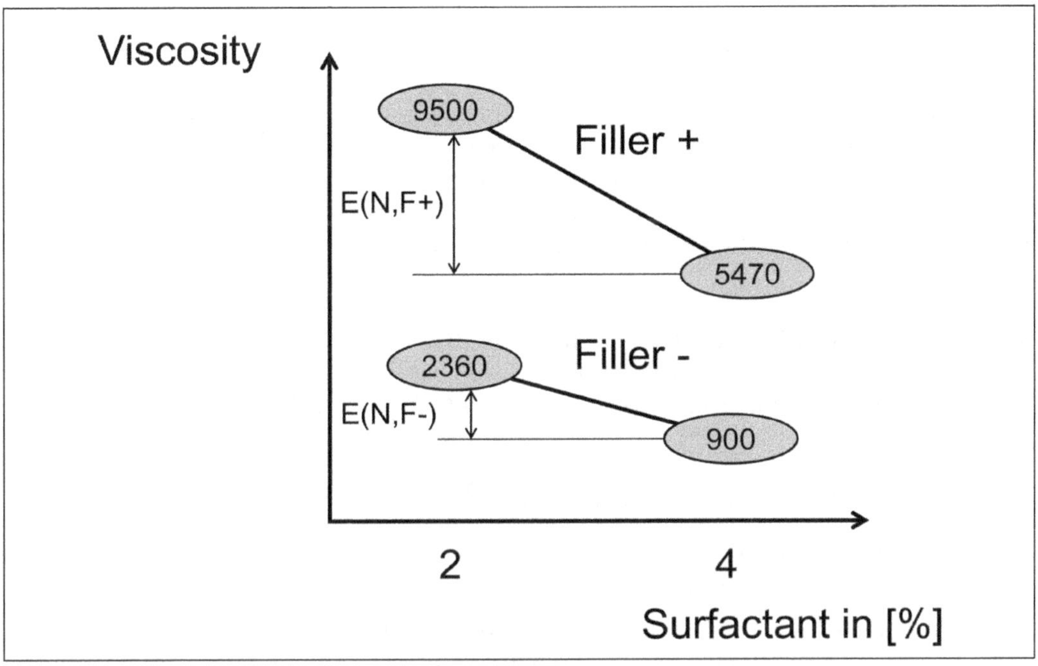

Figure 1.19: Interaction diagram for the example filler concentrate (viscosity values in mPas)

Interactions are also plotted in so called **interaction diagrams** (see Figure 1.19, 2.19 or 2.24). This diagram plots also the single effects, which assist during the data interpretation.

Interaction can occur in a different manner (see Figure 1.20). Normally both single effects have the same prefix as the main effect **(ordinary interaction)**. In case of strong interactions it is possible, that the single effects cross each other **(disordinary interaction)**. In this case, the value of interaction is larger than that of the main effect. Thus, a clear interpretation of the main effects is not possible and statements of the main effects are irrelevant.

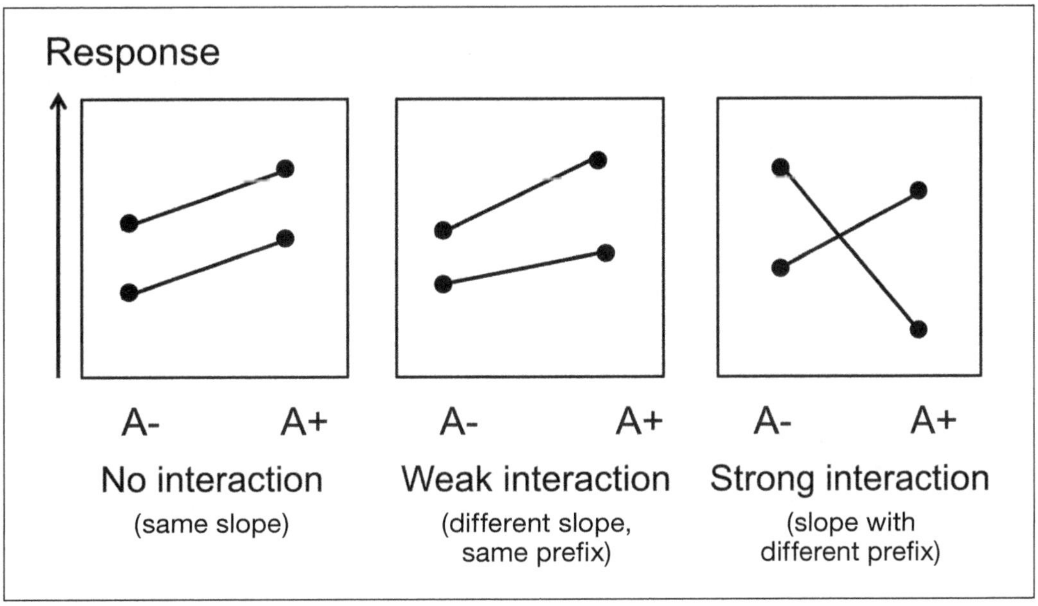

Figure 1.20: Types of interactions

In such situations it is advisable to look on the effects in diagrams. During dealing with DoE, the interactions should always be smaller than the main effects. Otherwise the main effects cannot be interpreted and analyzed apart from the other effects. Normally interactions - especially higher ones like ABC interactions (see Chapter 2.2.2) are small, because it is not very probable that the effect of a factor depends on the adjustment of three or more other factors. In case of strong interactions statements and interpretations about main effects must be modified, because they do not represent the reality. In these cases it is suggested to increase the distance between the levels, to clarify if the factor has a significant effect or not.

▶ ***Basis principle:*** Main effects must be larger than interactions. Otherwise they cannot be clearly allocated for interpretation. If this is not the case it is suggested to increase the investigation range and allow the effect of each factor to be captured.

1.6 Where is the statistics?

From the beginning of this book we talk always about a statistical approach in design of experiments. But, where is statistics? Until now it was not present.

To achieve the desired accuracy in data interpretation it is necessary, that multiple replications of the measurements are performed. The presence of a certain variance in the data cannot be avoided. These errors are the slight variations that occur when successive measurements are made by the same person under as nearly identical conditions as possible. However, omnipresent not controllable undesirables (noise - fluctuations in materials, surrounding conditions, the measurement, the execution of the experiments, etc.) will always influence the result. Considering the example with the filler concentrate it is obvious, that a certain fluctuation in the data of multiple formulated (replications) concentrates will always occur.

▶ ***Basis principle*:** Replicate measurements are normally associated with very small random errors, the noise. It's very unlikely, that replicates show identical average values, despite being obtained under identical conditions.

For the set of data in Table 1.5 following characteristic parameters can be calculated (see Appendix 2 and 4):

Arithmetic mean $\bar{x}$ = 3425 mPas
Standard deviation s = 386 mPas
Confidence interval VB (95 %, f = 3) = 614 mPas

The average viscosity of the filler concentrate has to be declared by:
Viscosity = 3425 +/- 614 mPas (at a shear rate of 100 s^{-1}, 20 °C)

Table 1.5: Viscosity of the filler concentrate in case of a multiple formulation (N = 4) at the data point (1) with 2 % surfactant and 20 % filler

Viscosity in [mPas]
2900
3600
3800
3400

This confidence interval is in the simplest case equal in the whole experimental area. Otherwise an average from all variances (= square of the standard deviation s, s^2, see Appendix 2 and 4) of each combination of factors has to be calculated by the geometric mean (see Appendix 2).

What is the sense of a **confidence interval** VB in case of the average viscosity? This interval defines a region in which the true value (see Appendix 3) of the viscosity can be found with a certain probability (in this case 95 %) (see Figure 1.21). There is no absolute security for that, because the region has to be calculated (or better estimated) on the base of only 4 values. To gain 100 % security in the ideal case an endless, at least a huge number of experiments has to be collected. This is due to time and cost aspects not possible. Thus, a control sample of 4 runs has been analyzed. The calculation of the confidence interval gives the possibility to implicate what will happen if endless runs will be done. At least we can come up, that the calculated average value will be in case of 95 data points from totally 100 inside this interval. Only in case of 5 experiments the value will be outside the predicted area (see also Appendix 4 and 5, t-test).

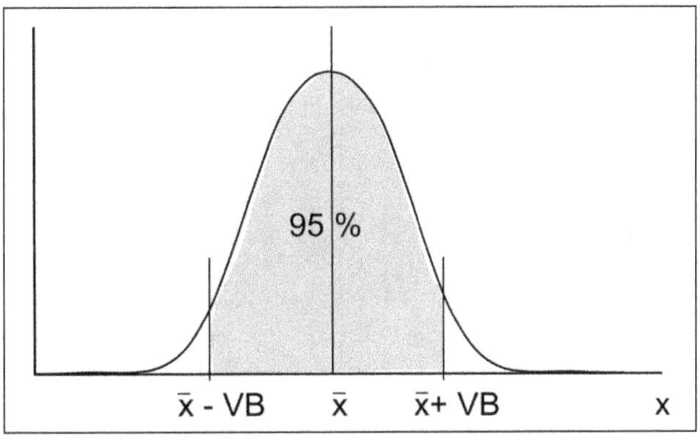

Figure 1.21: General plot of the confidence interval VB with a probability of 95 %

▶ ***Basis principle:*** The confidence interval defines a region in which the true value of the viscosity can be found with a certain probability (e.g. 95 %).

What are the consequences for data analysis in case of DoE? We have to consider these fluctuations! Every statement about effects on the viscosity has taken into account these fluctuations. In other words, not every difference between two values is an effect. The difference between two values is only significant (und therefore an effect can be stated), when the averages are separated by at least two times of the value of the confidence interval (see Figure 1.22). Otherwise the viscosity intervals of the two average values overlap and cannot be differentiated with 95 % probability. In other words, the probability that the difference between the two average values is only random is with 5 % very small.

In case of the filler concentrate a viscosity of 3000 mPas cannot be distinguished from a concentrate with 3100 mPas, because each value of viscosity has a fluctuation of +/- 614 mPas. The formulation and afterwards the viscosity measurement allow no more precise statement. Two average values need a difference of at least 2 · 614 mPas to be sure with 95 %, that there is a difference. In this context very often the terminus **confidence interval or the standard deviation of the main effects** can be found in literature (see Appendix 4). This uncertainty has to be taken into account already during the target formulation. Thus, the target viscosity 2000 mPas is fulfilled, when the value is inside the interval 3425 +/- 614 mPas (at a shear rate of 100 s^{-1} at 20 °C).

This uncertainty is taken into account during the data analysis and is plotted also in the so called ***Pareto*-diagram** (see Figure 1.23). This is a column diagram with the absolute values (is the non-negative distance to zero) of each main effect and interaction. Mostly the values are classified and the prefixes are also shown in the diagram. The limit of the confidence interval is plotted as a line. In most cases the double value of the standard deviation (2s), which is close to the confidence value with 95 % probability) is plotted. Dealing with many variables the geometric mean of the standard deviation of all main effects and interactions is favoured. Beside this way of illustration several other possibilities, which are described in Chapter 3.2 can be found. Very often a relative scale is used with so called **standardizes effects.** The effect will be divided by the standard deviation (see Figure 2.17).

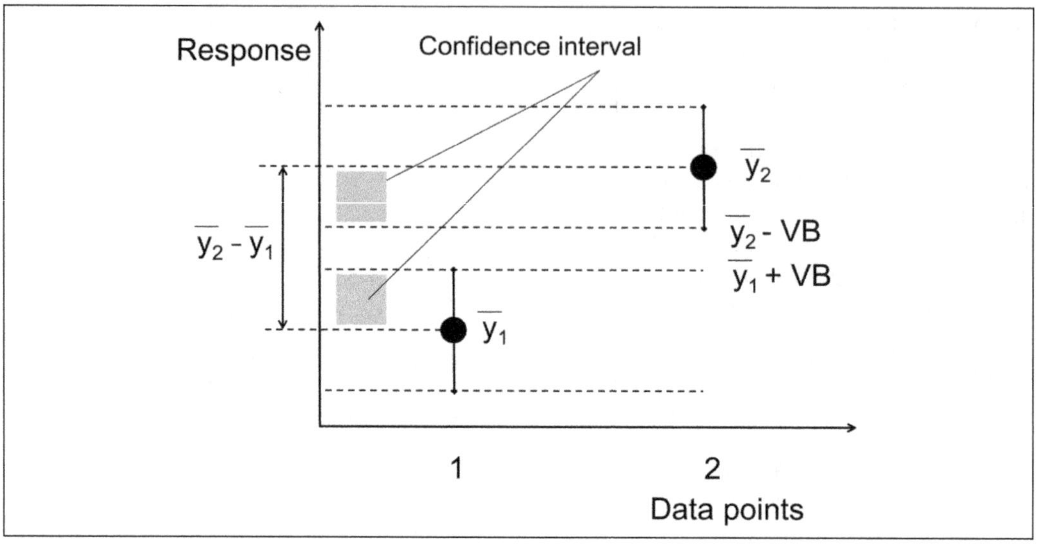

Figure 1.22: Uncertainty of statements in case of effects [18]

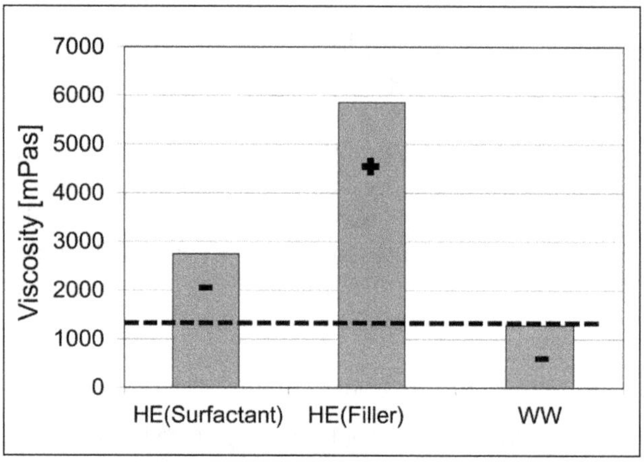

Figure 1.23: Pareto-diagram for the example for the filler concentrate. The dotted line represents the upper level of the confidence interval

Questions about the **significance of effects** are independent of all experimental design methods. It is anyway crucial for an efficient progress in projects. The target of a project can only be fulfilled in an efficient way, if the target value can be detected by an objective and validated method (the accuracy of the method is known, see Appendix 1 and 4). Unavoidable fluctuations and disturbances have to be taken into account already during target definition. Thus, discussions whether an alternative coating system has a better resistance as the reference are based on valid facts. The difference to the reference material has to be only compared with the precision of the measurement method given by the confidence interval - see Appendix 4. In many cases the confidence interval is larger than the supposed effect or the difference. Already *Aristotle* said: It is the mark of an educated man to look for precision in each class of things just so far as the nature of the subject admits. In such situations, system knowledge and experience is important to give statements of high quality. Also very often used reference materials or standards cannot help to interpret the effects. All results, also from the reference material, have the same fluctuations. However, such experiments with standards are valuable to check the plausibility of an experimental series! In addition, be aware, that a statistical significant effect is no assurance that it is large enough to have any practical value. Significance is no matter of blind trust. These principles are also relevant regarding technical specifications of products. The variance of the measuring method should consume only a certain amount of the tolerance limit of the product specification. Please consider for details regarding **measurement system analysis** further literature [38, 39].

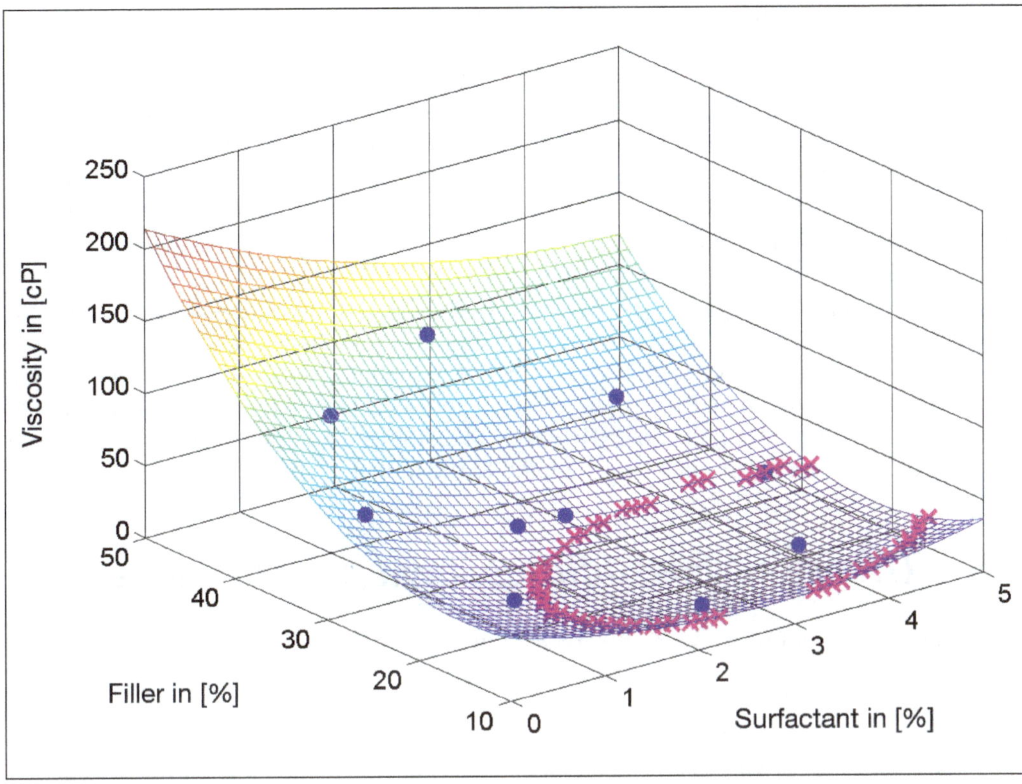

Figure 1.32: Quadratic model in case of the example of the filler concentrate including all 10 data points (dots) and an illustration of the optimal region regarding viscosity (crosses)

obtains slightly different model parameters. Also in this case an average value and a confidence interval of the model parameters can be calculated which contain the noise (see Chapter 3.3).

In case of the filler concentrate additional data points show, that even although the interaction is taken into account a substantial deviation from the linear model takes place. Finally the system is described by a quadratic model (see Figure 1.32). In addition in Figure 1.32 the area is marked, in which the target value of the viscosity is fulfilled. Experimental design for quadratic models will be described in Chapter 2.4 of the book.

Additional data points open up also the possibility to integrate additional parameters, like dispersing time or other parameters (e.g. sequence of the additive addition or the temperature). By this approach the last open questions about the influence of these parameters can be answered. DoE offers an easy and systematic approach for that, because the designs can always be expanded (see Chapter 2).

Thus, in contrast to the one-factor-at-a-time method design of experiments offers the calculation of a model. Therefore, the correlation between input factors and the response can be derived. **Models**

- compress date perfectly,
- opens up the possibility to make predictions about or simulations of certain situations without a big effort,
- show, which adjustment of the parameters ends up in an optimized operation point and
- show compromise solution in case of conflicting targets.

Thus, design of experiments opens great opportunities to optimize systems. A typical example for such an approach would be the reduction of raw material cost in parallel to keep the performance at the same level or even an increase in performance. In analogy to additional factors also additional target values (e.g. transparency) can be included. Thus, in the already identified area with filler concentrates which have an optimized viscosity, the recipe with the highest transparency can be identified. This approach can also be used for further measurements like storage stability of the viscosity and is called multiple response optimizations (see Chapter 4.2).

▶ ***Basic principle:*** Regarding modelling some basic principles has to be taken into account. The generated design of experiments and the chosen model are intimately linked. Not every design can be used for every model. Be careful with very simple models based on two or three data points. There is a high chance to run the risk of overseeing non-linear behaviour. However, plausibility has to be checked always and the experimental design has to be adapted to the model.

1.8 Overview, possibilities, benefits and limits

Statistical experimental design analyzes systematically the effect of some factors on a response. The methodology joins already in the preliminary planning and is also present during data analysis. The experimental effort is adjusted to the request (maximal information with a minimal effort) and an optimized ratio between the amount of relevant information and the number of experiments is realized. As a consequence time to market is reduced and also the R&D costs. With the global aim to improve the system knowledge, with the aim to make an optimization or even a preventive analysis of influencing factors and disturbances to avoid errors and design robust products and processes, DoE may increase at the project start the number of experiments in contrast to the classical approach. Especially in case of unknown systems in both approaches an intuitive experimentation is necessary to get some experience, to know the limits, etc. Without this knowledge it would be impossible to structure and define the questions for a selective application of the DoE-tool. On the other hand during data analysis information is compressed into a model. This describes how to affect the response by each factor. Thus precise and quantitative predictions about correlations in the system are possible and optimizing steps can be initiated. Finally this ensures the sustainable success with robust products and processes that can be widely used. Problems can be solved very fast and there is a perfect base for further developments (e.g. with a platform strategy). Already the huge and for all people available and documented system knowledge based on facts is of inestimable value. Thus, very often a considerable effort has to be overcome. However, in the long term this gives the best return.

Design of experiments can also be understood as philosophy, which is as package highly cross-linked with supplemental up- and downstream tools like project management, quality management, benchmarking, Quality Function Development (QFD), Failure Mode and Effects Analysis (FMEA), platform strategy, etc. Thus, design of experiments is an integral element of quality planning and quality management in R&D, construction and process planning. Finally, this element perfectly suits to the principles of lean-development [37, 38]. This mindset has the potential to revolutionize product and process design in many kinds. Due to planning the odds of success are increasing. Planning reduced the reaction time in case of external occurrences and there is a learning effect for the next time. However, planning can never predict the future, but its improving the ability to estimate certain aspects and to react in case of changes much faster and flexible.

What's possible with design of experiments and how do coating formulators benefit (**reality**)?

- More precise information is acquired in fewer experiments
- Identification of the significant influencing factors and interactions
- The design of experiments approach is balanced and objective, no combination has more importance than another
- Optimized statistical protection of the experiments
- Clear, precise and well defined targets
- Graphical illustrations lighten up the data analysis and documentation
- Avoiding the principle "just try" by a systematic and organized approach in a rational manner
- Realization of an empirical model
- Ensuring robust processes and products
- The experimental effort can be estimated and calculated before starting the experiments
- Results are evaluated in the light of variability and prevent wrong decision-making
- Already in case of conflicting targets an optimum can be found, because an extensive system knowledge is available
- Is a consistent way of thinking

What's not possible with design of experiments (**myth or legend**)?

- To replace the health common sense
- Generate creative ideas
- Replace knowledge and experience (e.g. experimental area)
- Balancing absent or wrong measurements
- To describe why a factor is effective
- To have a concrete recipe of success, especially without any modification for all problems

DoE is no culture to develop something automatically *(„Ars Inveniendi", G. W. Leibniz)* and no substitute for thinking about the problem. Thus, this method limits creativity in no way. Rather scientists get an essential tool, which can help them to work more efficient and to escape from the daily mania in the laboratory and the seeking of the needle in the haystack. The method is especially efficient for the analysis of several influencing factors on a few targets. Only the described DoE method can conserve in such situations the overview.

▶ ***Basic principle:*** Design of experiments cannot replace the health common sense or intuition and is not a package free of care. Using experimental design is no substitute for thinking about the problem and limits creativity in no way! Still an intensive analysis of the problem at all levels, also beside statistics (practical experience, theoretical background knowledge, etc.), is necessary.

Chapter 1 familiarize with elements of experimental design in general and DoE in detail. Table 1.7 once more arrange the relevant steps in DoE and assign the issues to the relevant book chapters. Some aspects has already been illuminated, other will follow in Chapter 2 and the following ones. Thus, this table assists also during orientation in the book and is a guideline in case of problems in practice.

Table 1.7: Overview about the different phases and steps in DoE and data analysis

Project phase	Activities	Methods	Chapter	Main message
Problem-formulation	• What's the problem? What should be achieved? • Formulation of an aim/ target (reproducable and measureable) • Covering the relevant factors and disturbance factors	• Brainstorming • Literature search • Structuring and classification • Top-Down-approach • FMEA • Mind-Mapping	1.1, 2.1 and Appendix 8	It is much more important to do the right things, than to do things right
Screening-design	• Screening for 20 % of the factors with 80 % of the effect • Find out the limits of the experimental design • Selection of the levels • Choosing the design (responsible of the model)	• Blocking • Randomizing • Normalize centering • Orthogonalize • Central points • Factorial or fractional designs • Design for non-linear models • Computer generated optimal designs	1.5, 2	If I have not very much time, I will take much of it at the very beginning
Experiments	• Running the planned experimental series	-	-	A good plan is the best background for improvisation
Analysis of the raw data	• Check if the normal distribution is valid • Reproducability • Time effects	• Normal probability plot • Outliers • Transformation • Blocking	1.6, 3.1, 3.3, Appendix 3	Nothing will come out of nothing
Data analysis	• *Pareto*-Diagram • Confidence intervals • Choosing a model (linear, quadratic, etc.) • Quality of the model, Check plausability	• Confidence interval • Regression analysis (correlation coefficien- observed vs. predicted, analysis of the residuals) • ANOVA • Response surfaces	3.2, 3.3, 3.4	Always think about the plausibility of number crunching
Optimization	• Adapting the model with additional data points • Analysis of all other effects • Multiple target optimization	• Optimization strategy like Simplex • Method of the steepest ascent • EVOP • evolutionary algorithms	1.7, 2.1, 4.1, 4.2	Finding a needle in the haystack
Robustness studies	• Finding the operating point where the response is as much as possible independent of disturbances	• *Taguchi* method • Analysis of the regression model	4.3	Errors which never happens, has not to be corrected
Useing				The project is ready, when all lacks are eliminated and the documentation is complete

1.9 A brief history of statistical design of experiments

The roots of DoE can be found in 1926. *Sir Ronald Aylmer Fisher* (British statistician, 1890 - 1962) developed in the 1930s the essential basics of factorial design, to improve the output in agriculture [9-11]. In this context also the analysis of variance as technique in data analysis has been drafted. In the 1950s and 1960s *George Edward Pelham Box* (British chemist and statistician, 1919) and *K. B. Wilson* created special experimental designs for optimization approaches and opened a huge field of applications in agriculture. They introduced also the term of the response surface. During this time *William Sealey Gosset* (Chemist, 1876 - 1937) established in addition the Student t-Test [12]. *Jack Carl Kiefer* (US-American statistician, 1924 - 1981) and *Jacob Wolfowitz* (US-American statistician, 1910 - 1981) analyzed in the same period the mathematical basics of optimal experimental design and developed the D-optimal design. After the Second World War the ideas has been exported to Japan and were integrated in the industrial research and development. This is especially also the remuneration of *Genichi Taguchi* (1924), who generated in addition the robust design philosophy of quality [17]. In the 1980s the western world considered the success of Japan and tried to relaunch the idea of DoE as well as some improvements has been performed [16]. In this context mainly *Dorian Shainin* (US-American engineer, 1914 - 2000) hast to be mentioned, because of his contributions in the fields of industrial problem solving, product reliability, and quality engineering [14, 15]. In parallel, optimization methods like Simplex (1962) and Evolutionary strategies (1979) have been developed. An important milestone for the extensive application of DoE in industry was the development of some commercial available software packages. Due to this step, it is no longer incredibly difficult to do the number-crunching.

1.10 References

General literature references can be found in the Appendix 9.

[1] S. Sorvia, Chemie Ingenieur Technik 1996, 68, 71
[2] B. Jandeleit et al., Angew. Chemie 1999, 111, 2648
[3] H. J. Bullinger, Einführung in das Technologiemanagement - Modelle, Methoden, Praxisbeispiele, Teubner, Stuttgart 1994
[4] http://www.expedition-bewusstsein.de
[5] T. Brinz, T. Burk, R. Emmerich, J. Ortmeier, Farbe & Lack 2009, 115 (12), 32
[6] J. Schmelzer, Farbe & Lack 2003, 109 (12), 22
[7] R. Koch, Das 80/20-Prinzip. Mehr Erfolg mit weniger Aufwand, Frankfurt/M., New York, 1998
[8] http://www.statistics4u.info/fundstat germ/cc modeling.html
[9] R. A. Fisher, The arrangement of field experiments, Journal of the Ministry of Agriculture of Great Britain 1926, 33, 503
[10] R. A. Fisher, Statistical Methods for Research Workers, Oliver and Boyd, Edinburgh, 1925
[11] R. A. Fisher, The Design of Experiments, Oliver and Boyd, London, 1966
[12] Student, Biometrika VI, 1908, 1, 1
[13] C. D. Montgomery, Design and Analysis of Experiments, 7th Edition, Wiley & Sons, 2008
[14] D. Shainin, P. Shainin, Quality and Reliability Engineering International 1988, 4, 143
[15] K. R. Bhote, Qualität - Der Weg zur Weltspitze, IQM, Großbottwar, 1990
[16] B. Gunther, A Perspective on the Taguchi Methods, Quality Progress, 1987, 44
[17] G. Taguchi, System of Experimental Design, Vol. I + II, Kraus International Publications, New York, 1987
[18] M. Adam, Statistische Versuchsplanung und Auswertung, Düsseldorf, http://mv.fh-duesseldorf.de, 2010
[19] T. Tietjen, D. H. Müller, FMEA Praxis, Verlag Hanser München, 2003, p. 41
[20] K. Götz, Integrierte Produktentwicklung durch Value Management, Shaker Verlag, Aachen, 2007
[21] VDI-Gesellschaft Produkt- und Prozessges. (Herausgeber), Wertanalyse - das Tool im Value Management, Springer, Berlin Heidelberg, 6th Edition, 2011
[22] L. Braun, C. Morgenstern, M. Radeck, Prozessoptimierung mit statistischen Verfahren, Hanser Verlag, München, Wien, 2010

[23] T. Pfeifer, Qualitätsmanagement, Hanser Verlag, München, Wien, 2. Auflage, 1996
[24] B. Jandeleit, D. J. Schaefer, T. S. Powers, H. W. Turner, W. H. Weinberg, Angew. Chem. 1999, 111, 2648
[25] J. Ortmeier, S. Ansorge, H. Kost, Farbe und Lack, 2007, 113, 35
[26] U. S. Schubert, besser lackieren, 2003, 14
[27] G. Schuh, Produktkomplexität managen, Hanser Verlag, München, Wien, 2 Auflage, 2005
[28] S. Thomke, D. Reinertsen, Die sechs Mythen der Produktentwicklung, Harvard Business Manager 2012, Juli, p. 68
[29] R. Gunther McGrath, T. Kell, Gescheiterte Projekte profitabel vermarkten, Harvard Business Manager 2007, September
[30] A. Seiler, Planning, BWL in der Praxis III, orell füssli Verlag AG, Zürich, 2000
[31] A. Schröder, F&E-Manager 2010, 1, 6
[32] J. Kuster, et al., Handbuch Projektmanagement, Springer Verlag, Heidelberg, 3. Auflage 2011
[33] R. Baumstark, B. Lohmeijer, Let's Talk Wood 3, BASF Event, Berlin 2012
[34] S. C. Wheelwright, K. B. Clark, Revolution der Produktentwicklung, Neue Zürcher Zeitung, Frankfurt/Main, 1994
[35] M. H. Meyer, A.P. Lehnerd, The Power of Product Platforms, Free Press, New York, 1997
[36] M. E. McGrath, Product Strategy for High Technology Companies, McGraw-Hill, New York, 1995
[37] J. M. Morgan, J. K. Liker, The Toyota Product Development System - Integrating People, Process, and Technology, Productivity Press, New York, 2006
[38] A. C. Ward, Lean Product and Process Development, Lean Enterprise Institute, US, 2007
[39] C.P. Ceferstein, Fertigungsmesstechnik, 7. Auflage, Vieweg + Teubner Verlag, Springer Fachmedien, Wiesbaden, 2011
[40] L. Braun, C. Morgenstern, M. Radeck, Prozessoptimierung mit statistischen Verfahren, München, 2010
[41] G. Apitz, Workshop Modulare Technologie, Aschaffenburg, November 2012

2 Planning is essential - a lot helps a lot

2.1 General principles for setting up a DoE investigation

2.1.1 Strategy of experimentation and guidelines for pre-experimental planning

One of the essential success factors in research and development is planning of the experiments (see Chapter 1). Very often people invest not enough time into this phase of a project and/or work only with insufficient accuracy. This may end up in considerable wrong decisions and costs. Planning is the base for all subsequent steps. Naturally remarkable potential savings are available by this front loading of the effort. Hence, planning has almost more importance than data analysis and a lot helps a lot. In other words as the psychologist *Ruth C. Cohn* said: If I have not much time, I take plenty of that at the beginning.

Basically **three important stages** in the whole DoE-process (see Table 1.7) can be distinguished:

- **Pre-experimental planning:** At the beginning – as already emphasized in Chapter 1 – a **system analysis** should take place.
- **Screening:** In this step a **preliminary ranking of the system parameters** takes place and the most important ones are identified with a minimal amount of experiments. Thus, we start with a rough overview of the system and separate the parameters of no importance based on the *Pareto*-principle considered in Chapter 1.4.2.5. In these first experiments, 20 % of the factors should be identified, which are responsible for 80 % of the effects. All other factors should not be studied before the significant main parameters are known. This should be done in a second step by fixing the main parameters on a certain level (**held-constant factors**) and varying the other influencing factors. Normally processes are too complex to generate adequate system knowledge already in one step. In addition, the experimental setup would not be economically practicable. Therefore the design should be kept as simple as possible.
- **Modeling/optimization/robustness testing:** In this step (very often also identified as response surface phase) the precise influence of each single factor on the target is determined. A model is fitted to the data (linear response, interactions, quadratic terms, etc.), which open up the possibility for predictions and optimizations. Thereby a problem should be described as simple as possible, but not simpler. To get a precise statement, the list of influencing factors must be short in this phase. This is due to a high number of experiments because of the already small effects.

Experimental designs will depend on the DoE-stage. However, a so called **pre-experimental planning** stage is important to avoid that DoE fail due to insufficient necessary preparations. In this step, essential questions have to be answered:

- **Which problem should be solved by the experiments – recognition?** It is crucial for the project success, that a clear formulation of the problem and the aim (why is an experiment done, what's the purpose and what is the desired result?) is accomplished on transparent data, numbers and facts. A vague conception of the aims in projects is a no-go situation. This part is commonly underestimated and less time is invested into this duty. Very often, people start to work on base of the first perceptions of the problem, which sketch a huge, vague and complex issue. However, it's very important to catch the situation and denote the real subject of the problem. The **5 Whys-method,** an integral content of the success story of *Toyota* might be helpful in this step. This iterative question-asking technique is used to explore the cause-and-effect relationships underlying a particular problem. The primary goal of the technique is to determine the (initiating) root cause of a defect or problem [9].

- Already in this first period a **structured approach** is meaningful. On one hand, the problem should be simplified and structured, as well as the scope defined. Typical questions are: Which steps of a process are connected and succeed chronologically? Which information is essential for each step? Very often, the project is split into sub-projects. On the other hand, it has to be imposed which information and effects are crucial for the question. Aspects, which are not taken into account in this step, will end up later on in lacks of the cause and effect correlations (model).
- **Who is concerned by the project?** It's of best practice to involve all concerned people. This is especially important for the definition of the aim and during the system analysis (different experiences can be taken into account). If there is no clear common purpose, each participant follows his own individual and concealed aim. Stress, conflicts and inefficiency are inevitable due to the not invented here syndrome. Be aware of the false conclusion, that all others in the team have the same picture in mind as you have. A team approach with solicit input from all concerned parties is strongly recommended.
- **What is already known about the present problem – familiarization?** Background of each planning is information. The allocation of all already done experiments and investigations as well as present literature, physical theory, etc. is crucial to familiarize with an issue. Knowledge beside statistics is invaluable in choosing factors, factor levels, the number of replicates and data interpretation. The wheel has not to be invented twice. However, if the researcher is already familiarized with the application, one may skip this step and go directly to the next one. On the other hand, if no information is available, one should proceed to exert some kind of influence on the investigated system. In such situations it is desirable to find out whether the process is possible at all and roughly how it goes. This may be accomplished already by a few experiments.
- **Agreement about the target values/responses and the aim:** Primary, target values which will be used for measuring the outcome of an experiment, has to be clarified. An incorrect selection of the responses annuls all further research activities. In addition, a precise formulation of the aim must be present and all essential factors should be included, not only the current issues. Especially an intensive discussion about the target values improves the quality of the subsequent experiments. Thus, the mentality of "just trying it" is disabled already in the front end. Choose always all responses that are relevant to the goals set up in the problem formulation. Very often even more than one response is necessary for that. Pay attention, that aims are measureable (e.g. scratch resistance should be below 1 N), because otherwise it is impossible to evaluate the quality of a project. Target values should also be determined reproducible. In this stage, we will therefore have to verify (**validation**) that the response measurement techniques work in reality. In this stage as much as possible of the disturbances or uncontrollable factors on the response measurement techniques (e.g. atmospheric conditions, trends like warming up of instruments, etc.) has to be taken into account. This enables to consider the unavoidable fluctuations and disturbances by tolerance limits already during setting the targets. Subsequent discussions can be moved down to solid basics pre-emptive. In many cases the variance is larger than the supposed effect or difference. In case of time series (e.g.

storage stability), it's very effective to observe the time-shift of some measurable values between the start and the end of the time period (e.g. change in viscosity, drying time, etc.). Time influenced changes of the target value and possible causes can be identified by this way. It is recommended to analyze also the real values of the target before and after the time period by *Pareto*-diagrams. Thus, it is guaranteed that additional temporal changes of the ratios between main effects and interactions can be identified.

- **Identification of the relevant and all other – the system characterizing – factors and noise factors.** It's important to collect all influencing factors, also those, which will be constant in the further experiments. This improves the reproducibility of the experiments. The best statistical method can never take into account not observed factors! On the other hand, it's important to find a balance between the number of factors and the experimental effort. Every additional effect increases the costs, so keep the design as simple as possible. Certainly a possible effect can only be observed, if a variation of the factors takes place. Sometimes it's helpful for the decision, to distinguish between easily (e.g. flow rate) and heavily changeable (e.g. temperature of a container) factors. Very often in practice, there is a focus on some single factors, which might be mainly responsible. However, such fixation will end up in disappointing solutions. It's beneficial to evaluate the relevant influencing factors by **FMEA (Failure Mode and Effects Analysis**, see Appendix 8). Technicians tend to consider everything as relevant and will be caught in details. This approach exceeds a simple brainstorming or the collection of the parameters in a table or a cause and effect diagram (*Ishikawa*-diagram). Only really relevant parameters are determined by this method. In addition, a cause-related structuring takes place. Details can be found in Appendix 8.
- A **problem should always be described as simple as possible**. To deal effectively with design of experiment, the number of factors should be small. 6 factors are perfect. However, the number should be at least below 12. Many other factors, which were meaningfully collected, should be fixed at a constant level. In addition, it's helpful, to **combine directly connected variables by a ratio** (i.e. the concentration of an active component and it's stabilizer as ratio active component/stabilizer). Some groups of very similar substances can be collected as group (e.g. sum of fillers $A_1 + A_2 + A_3$).

▶ ***Basic principle:*** The basis for decisions of every experimental design is a precise formulation of the aim during the pre-experimental planning. What should be achieved and what kind of investigation is required? Afterwards it is important to evaluate all influencing factors, which might influence the target value. The best statistical approach can only describe factors, which are also under examination. Afterwards, the determination of those 20 % of parameters, which are responsible for 80 % of the effects (*Pareto*-principle) is recommended in a preliminary screening step. Otherwise the experimental effort will be too large, especially in case of complex processes. In addition, not all parameters have to be varied simultaneously. Keep the design as simple as possible and focus on the relevant influencing factors. Details are analyzed in a second optimization step, with a focus on only a few factors (Table 1.7).

Accepted aims and influencing factors as well as noise factors are the result of the creative preliminary analysis. Afterward it is necessary to define the next steps (Table 1.7):

- **Determine the levels and the range of the factors (adjustment of the data points)** adapted from well known aspects, literature or even intuition. Very often pilot experiments help to clarify. Especially in case of poor knowledge about the experimental area **preliminary tests** has to be performed. Each point in the experimental region has to be feasible. Thus, it is necessary to identify **restrictions** in the front end. In case of technical limits (e.g. unwanted process complications, excessive costs, highly toxic by-products, etc.) a certain distance to extreme values makes sense. In such cases the factors might shrink or irregular designs are used as described in Chapter 2.1.5. However, it is crucial to make the investigation range large enough to allow the effect of each factor to be captured. Large effects should be always analyzed to avoid disturbances due to noise. An improved reproducibility due to more replicates may help, but this will end up in more experiments. In general, the less one knows about a system (e.g. screening), the longer the intervals. Otherwise it is not certain beforehand where the best settings are found and to cover the entire range of reasonable values is covered ("**bold experimentation**"). In the optimization step it is possible to narrow the ranges. However, be aware that the experimental region is not too large. If the levels are too far from each other, variations from linearity (e.g. minima, maxima) may be considered not enough. In addition, designs with two levels may overlook non-linearity and sometimes extreme values like maxima or minima will be undetected an adulterate the results.

- **Specify the number of levels**: If many effects are included into the experimental design, only two levels are used in the screening period, to evaluate, which effects are significant. In subsequent phases the number can be enlarged. Very often, especially by qualitative effects, the number of levels can be derived from the question itself. A pump can only be switched on or off.
- **Maximal number of data points**: How many time and money is available for the experiments and which level of precision is necessary for the model and the explanatory power? These questions have to be answered before the maximal number of experiments can be defined. Be aware of some replicate measurements for reproducability reasons (see Chapter 2.1.2 and 2.2.4).
- **Choosing** the combinations of factors and steps – the **experimental design**. In principle the design depends on the model. Thus, also this aspect has to be taken into account. If it is clear from the very beginning, that a non-linear system is present, only adequate designs for non-linear systems should be chosen (see Figure 1.31). For components with concentration < 5 % (e.g. additives) usually a **factorial design** can be used. Variations in the concentration can be balanced by other components (e.g. solvent) without changing the system. For higher concentrations > 5 % (e.g. main components like binders) **mixing designs** (see Chapter 2.5) should be chosen. A variation of the content cannot be done without a drastic influence on the total system. This is also valid, if a maximal concentration of a group of components is an initial condition. In this case the sum of components will be used as effect. The other parameters will be integrated as quotient $A_i/(A_1 + A_2 + \ldots + A_i)$. In factorial designs no variations of small effects should take place in parallel to variations of the main components. In this case, it is better to create a mixture design for the main components and a subsequent variation of the other parameters with fixed main components (see Chapter 2.5).
- The **sequence of the experiments** may be important if time effects (e.g. exhaustion, warming-up of the equipment, etc.) are present.
- Clarify **responsibilities and deadlines**.

▶ ***Basic principle:*** To deal with DoE efficiently, the number of factors should be below 6, maximal 12. Many other factors, which were meaningfully collected, should be fixed at a constant level. It's important to collect all factors, otherwise the reproducibility will be seriously affected. The border of the experimental design has to be identified in the front end. Data points should not be too close together. A small distance may end up in not significant effects due to the variance of the data. On the other hand, if the levels are too far from each other, non-linear behaviour occurs frequently and extreme values like maxima or minima might be overlooked. The aim is to find correlations between factors and the target value (linear, interactions, quadratic terms, etc.). Probably present noise factors have to be identified and minimized if possible. The experimental design is correlated with the chosen model.

2.1.2 Overcome experimental errors – identical replication

Despite performing replicates of experiments under the same overall conditions with enormous care, results will never end up in the same values (see Chapter 1.6). **Identical replicates** improve the valid of each single result and statistical tools can give also information about the variance. However, each identical replicate increases the number of experiments. On the other hand variance is diminished. Thus, a balanced ratio between effort and result is necessary, expressed by an increase in the precision of the predictions. Without any replicate it is impossible to get an estimation of the variance.

In case of 2 or 3 influencing factors one replicate in each data point is enough. In case of more than 4 influencing factors it is recommended to measure between 3 to 10 replicates (usually in the centre point, see Chapter 2.2.4).

To get a complete picture, it should be mentioned that also methods are available, which can deal with experimental designs without any replicate measurements. In this case variance is estimated from random effects and no longer from the difference between replicates. Especially in case of very much possible influencing factors (more than 5) usually only a few are dominant. Thus, the majority of the effects will not be significant. These data can be used in a first estimation to calculate the variance (= **pooling**) [3].

▶ ***Basic principle:*** For a reliable determination the variance it is necessary to repeat the experiments (identical replication, reproducability).

2.1.3 How to overcome trends – randomization and arrangement in blocks

In addition to errors in measurements also other parameters may have an influence on the results. For example the quality of the raw materials may change or operating conditions may not be consistent over time (i.e. humidity will increase during the experiments, increasing soiling of the instruments or a warming up of the pilot plant takes place) and may end up in a trend of the experimental data. To prevent, that such an unknown time trend or other changes may distort the results, a random arrangement of the experiments is done (= **randomization**). The sequence can be derived from random numbers (flipping a coin, random generator, etc.). Afterwards, it is not allowed to change the sequence. Thus, randomization is not capriciousness and excludes only the mixing of effects, minimizes disturbing factors and therefore reduces the random error (see Table 2.1).

Table 2.1: Randomization vs. not randomized design [1]

	Not randomized			Randomized			
Number of the experiment	Factors			Factors			Number of the experiment
	A	B	C	A	B	C	
1	-1	-1	-1	-1	-1	+1	2
2	-1	-1	+1	+1	-1	-1	5
3	-1	+1	-1	+1	-1	+1	6
4	-1	+1	+1	+1	+1	+1	8
5	+1	-1	-1	-1	-1	-1	1
6	+1	-1	+1	-1	+1	-1	3
7	+1	+1	-1	+1	+1	-1	7
8	+1	+1	+1	-1	+1	+1	4

▶ ***Basic principle:*** To avoid trends and to be aware of any preferences regarding certain experimental situations, the allocation of data points has to be done by randomization. Randomizing all runs belonging to an experimental plan transforms systematic effects of uncontrolled factors into random, experimental noise.

However, randomizing may not be sufficient to deal with any undesired systematic variability that does occur. **Blocking** is another important method to reduce the variance of experimental data. In addition, a quantification of the influencing factors is possible. The basic principle is the collection of single experiments to groups of so called block and to perform not all experiments in a completely random order. Hence, the experimental error will be minimized by increasing the degree of similarity of the experiments in the group. In the block, each combination of factors and levels should occur in the same amount and noise factors should be constant. Thus, due to this balance, differences of the results between blocks can be identified, quantified and eliminated very easy (see Table 2.2).

Table 2.2: Blocking

Block	Factor A (temperature)	Factor B (amount of additive)
1	+1	+1
1	-1	-1
1	+1	-1
1	-1	+1
2	+1	+1
2	-1	-1
2	+1	-1
2	-1	+1

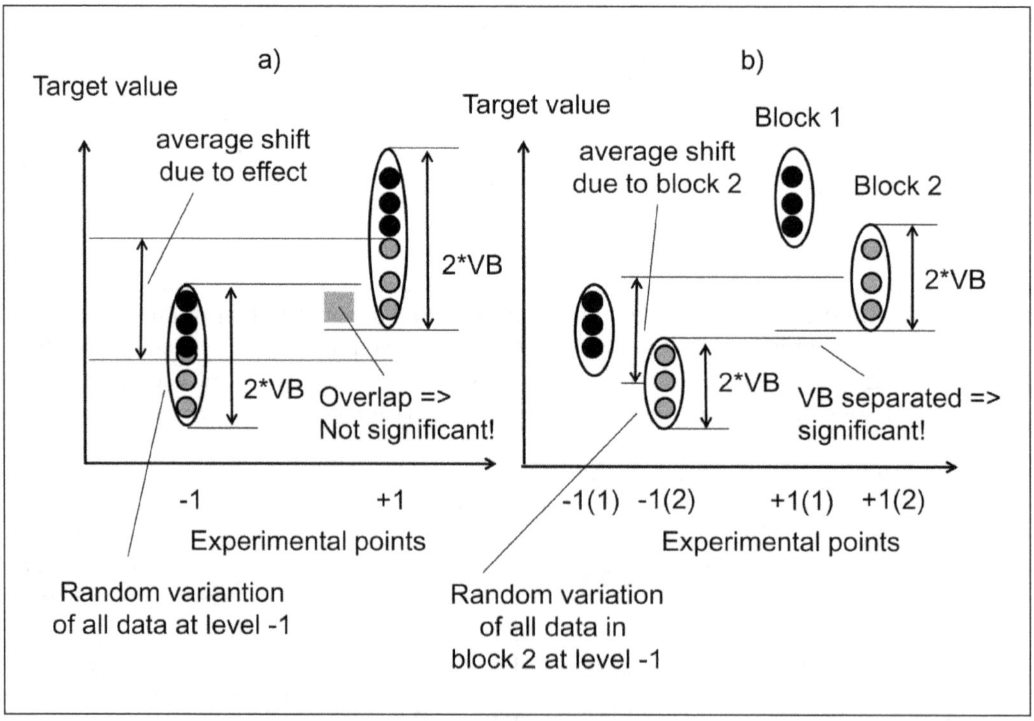

Figure 2.1: Effect due to blocking in case of a data set, which is based on two experimental series, which are not performed completely similarly. Due to the separation it is on one hand possible to identify a significant effect of the influencing factors and on the other hand also the difference between the two experimental series (block 2 is on a significant lower level of the target value). In parallel, the variance of the data is reduced to a large amount. If all data are analyzed as one data set, the effects become blurred in a larger confidence interval.

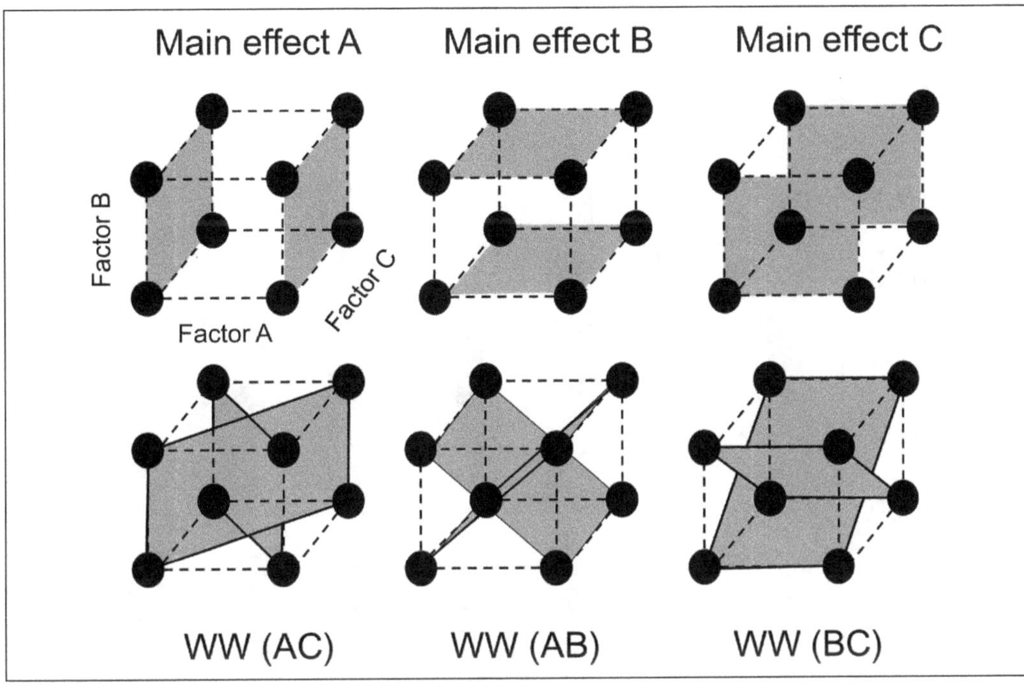

Figure 2.7: Geometric illustration of the main effects and interactions inside the cube for the example of the oscillatory-rheological measurement with an aqueous pigmented coating

If Equation 2.9 is written in another way around, it become more clear, that the interaction is the difference between the two average values from all data points on the diagonal-surfaces of the cube (see Figure 2.7).

Equation 2.10:

$$WW(V1V2) = \frac{E(V1,V2+,L-) + E(V1,V2+,L+)}{4} - \frac{E(V1,V2-,L-) + E(V1,V2-,L+)}{4}$$

$$= \frac{((1.24 - 1.14) + (1.43 - 1.22)) - ((1.14 - 1.17) + (1.25 - 1.12))}{4}$$

$$= \frac{1.24 + 1.43 + 1.12 + 1.17}{4} - \frac{1.14 + 1.22 + 1.14 + 1.25}{4} = 0.05$$

The two other two-factor interactions WW(LV1) and WW(LV2) can be calculated in the same way. In case of three influencing factors also a three-factor interaction is possible. If this is significant, the effect of one factor depends on the level of both other factors. In other words, the interaction of the two thickeners WW(V1V2) depends on the level of the solvent. Thus, all three influencing factors have to be taken into account together! This interaction is calculated from the average difference between both two-factor interactions ($WW_{L+}(V1V2)$ and $WW_{L-}(V1V2)$) at high and low level of the third factor (L). In practice, this interaction is rarely significant.

Equation 2.11:

$$WW(LV1V2) = \frac{WW_{L+}(V1,V2) - WW_{L-}(V_1,V_2)}{2}$$

$$= \frac{1/2\,(E(V1,V2+,L+) - E(V1,V2-,L+)) - 1/2\,(E(V1,V2+,L-) - E(V1,V2-,L-))}{2} = -0.01$$

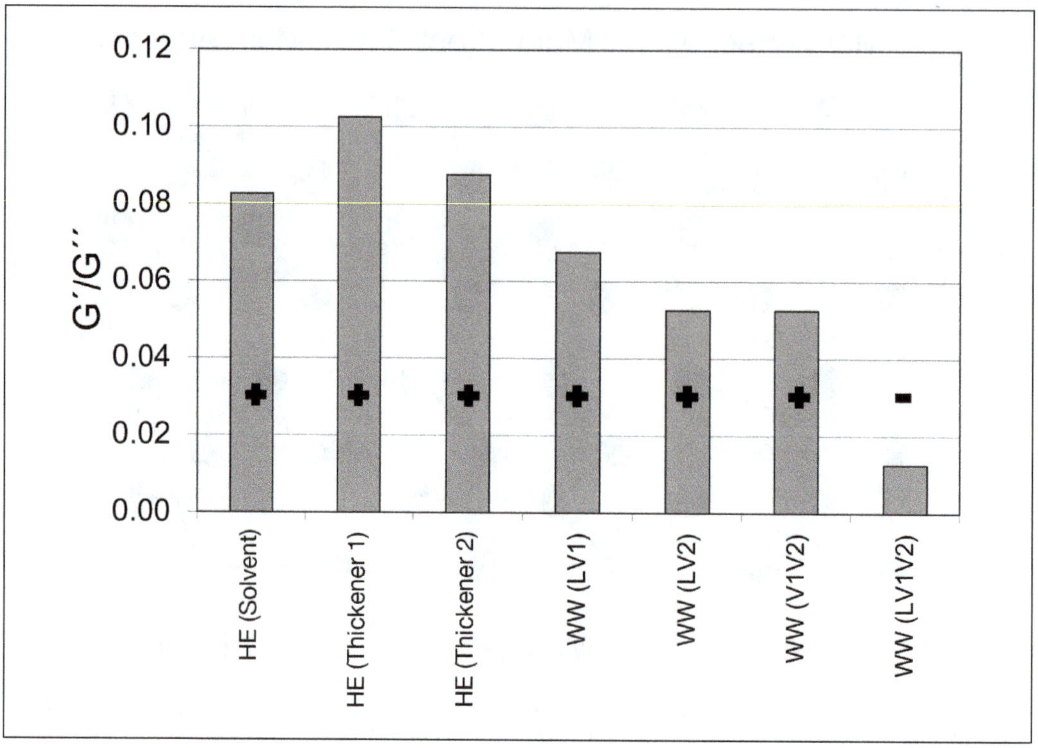

Figure 2.8: Pareto-diagram for the example of the oscillatory-rheological measurement with an aqueous pigmented coating

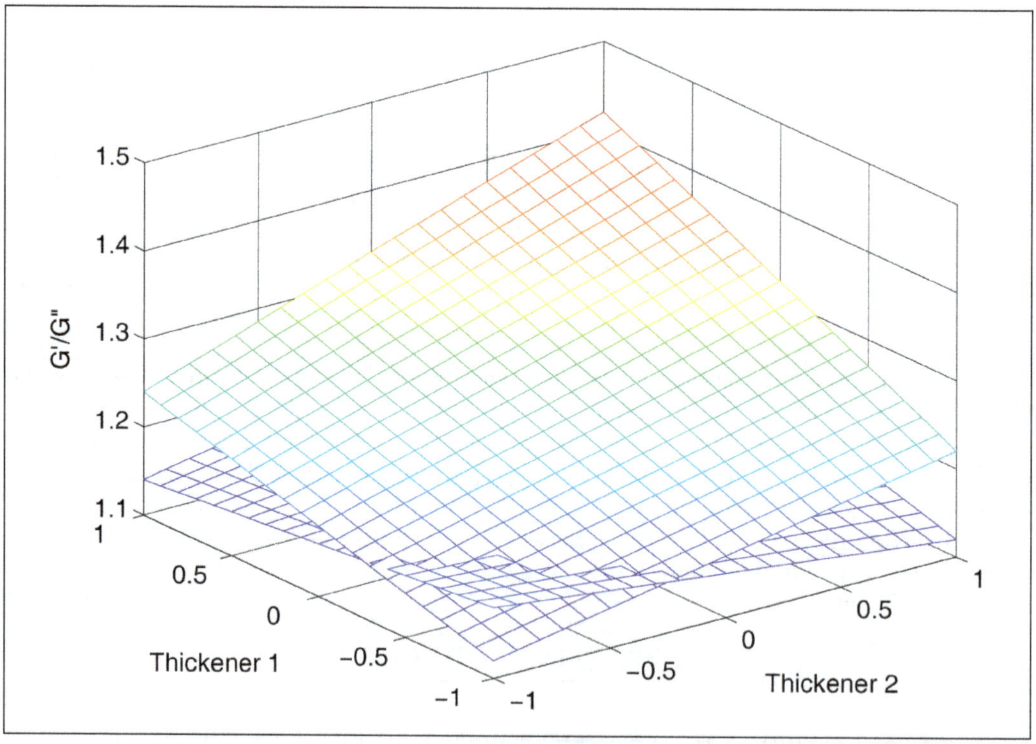

Figure 2.9: Response surface for the effect of both thickeners at high and low level of the solvent for the example of the oscillatory-rheological measurement with an aqueous pigmented coating

As already shown for the 2^2-design, all main effects and interactions can be graphically illustrated by *Pareto*-diagrams (see Figure 2.8).

Interpretation of the data:

- Thickener V1 has the highest effect on the ratio G`/ G``, followed by thickener V2 and the solvent L.
- All effects are positive, meaning that an increase in concentration of the additives end up in a higher ratio of G`/ G``.
- Clear interactions of all parameters are present. Thus the effect of one factor depends on the values of other factors.
- All interactions are smaller than the effects (this is an important prerequisite - see Chapter 1.5.2). The combination of thickener V1 and solvent L offer the most important synergistic interaction.
- The combination of all three additives offers a slight negative interaction. However, all important interactions are synergistic.

Table 2.8: Overview about the factorial designs in case of two levels = 2^k-Pläne

Number of factors	Label of the factorial two-level design	Number of experiments
2	2^2	4
3	2^3	8
4	2^4	16
5	2^5	32
6	2^6	64
7	2^7	128
8	2^8	256

To have an entire description of the system, a model with interactions can be derived from the main effects (see Chapter 1), which describes the system (G`/ G``= Thickener 1 · 0,050 + Thickener 2 · 0,045 + Solvent · 0,040 + Thickener 1 · Thickener 2 · 0,025). It is impossible to illustrate the effect of three variables directly in a graphic. However, it is possible to show the response surfaces on two levels of the third influencing factor (see Figure 2.9). The distance between the two surfaces represents the effect of the solvent. This way offers a possibility to illustrate clear the synergistic effect of the solvent on both thickeners. Thus, it is impossible to look at one of these parameters independent from the others. An increase in concentration ends up in case of all parameters in a higher ratio of G`/ G``. In other words, the elastic part (structure) gets more important. The addition of solvent increases also the structural amount due to an increasing interaction with the thickeners.

Table 2.9: Structure of a factorial design from simple building blocks

Number of experiment	2^4-plan			
	2^3-plan			
	2^2-plan			
	A	B	C	D
1	-	-	-	-
2	+	-	-	-
3	-	+	-	-
4	+	+	-	-
5	-	-	+	-
6	+	-	+	-
7	-	+	+	-
8	+	+	+	-
9	-	-	-	+
10	+	-	-	+
11	-	+	-	+
12	+	+	-	+
13	-	-	+	+
14	+	-	+	+
15	-	+	+	+
16	+	+	+	+
....				

2.2.3 *The general design with two levels - 2^k-design*

Experimental designs can be extended with additional influencing factors very simply - so called 2^k-design are produced (see Table 2.8). Higher designs always result from the combination of 2^2-design (see Table 2.9). Data evaluation is done in the same way as already shown. However, it is strongly rec-

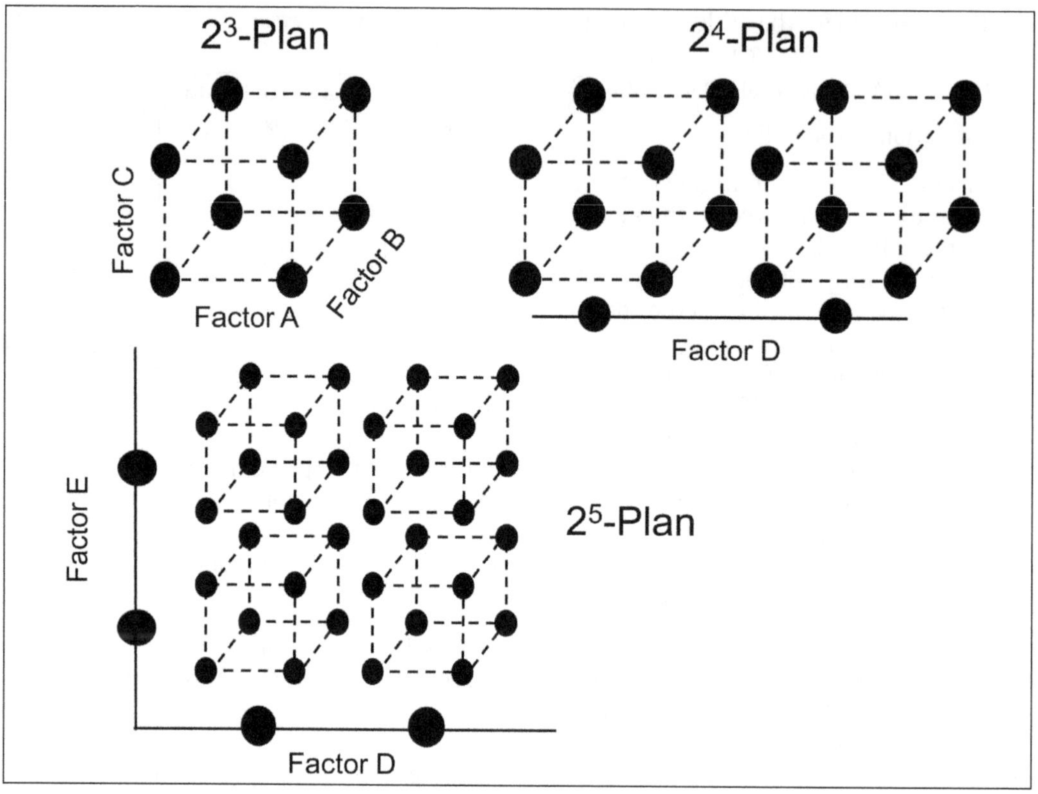

Figure 2.10: Design matrix with combination in case of a factorial 2^k-design for k influencing factors, each with two levels

ommended to use software tools (see Chapter 5). More than 3 influencing factors cannot be illustrated in one single figure, so other constructions are necessary (see Figure 2.10). In general the number of experiments can be calculated by Equation 2.12.

▶ ***Basic principle:*** Experimental designs can be extended with additional influencing factors very simply. So called 2^k-design are always the result from the combination of 2^2-designs.

2.2.4 Factorial designs with centre-points

A linear model can be very often accepted as a good simplification. To evaluate this assumption, usually a centre-point is added to the design (see Figure 2.11 and 2.12). The data value at the **centre-point** is afterwards compared with the calculated value from the model.

The centre-point is also perfectly suited for **replicate measurements**. The following rule of thumb can be applied [2]:

- In case of 2 or 3 influencing factors (thus, 4 or 8 data points) one replicate in each data point is enough
- In case of more than 4 influencing factors (thus above 16 data points) it is recommended to measure between 3 to 10 replicates in the centre point.

The number of replicates depends also on the experimental method. Procedures with low precision (see Appendix 1) like qualitative evaluations need much more replicates as methods with high reproducibility. This data point has no influence on main effects and

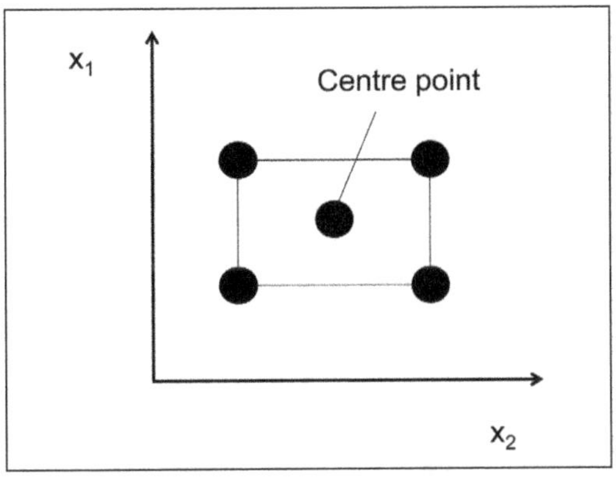

Figure 2.11: 2^2-Faktors with centre-point

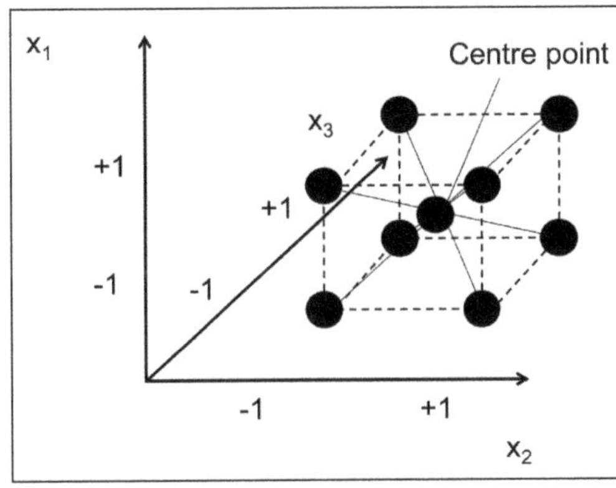

Figure 2.12: 2^3-Factors with centre-point

interactions, because acting as an average value of all experimental results. Thus, with a centre-point non-linearity cannot be allocated to certain factors.

▶ ***Basic principle:*** To evaluate the assumption of a linear model, in general a **centre-point** is added to the design. The centre-point is also perfectly suited for replicate measurements. As a rule of thumb one replicate in each data point is enough. The number of replicates depends also on the experimental method.

2.2.5 *Blocking with factorial designs*

Splitting of the experiments into identical blocks has been shown already in Chapter 2.1.3 as efficient tool to prevent distortion due to perhaps existing trends or other systematic differences. The following example shows, how data evaluation is done afterwards, is shown in the following example.

Example: Trouble shooting – discoloration of a clear coat, 2^4-factorial design with blocking

The reason for an unexpected discoloration of a clear coat directly at the customer site should be evaluated by design of experiments. Target value is the optical evaluation of the liquid coating in a glass beaker compared with a standardized grey scale. All data are measured twice. Four raw materials are relevant: matting agent, thickener and in-can preservative. Due to some public holidays, the experiments have to be done with one interruption within 14 days. To avoid distortion by time effects, the experiments are splited in two blocks. All replicate measurements are done on another day as the first run.

Properties of the experimental design (see Table 2.11):

- Two-level factorial design without randomization
- Number of factors: 3 (see Table 2.10)
- Number of blocks: 2
- Number of target values: 1
- Number of experiments: 16

Table 2.10: Factor settings in case of the example discoloration of a clear coat

Factors	Low limit in [%]	High limit in [%]	Levels
Matting agent	0	1.0	2
Thickener	0	0.8	2
In-can preservative	0	0.2	2

Table 2.11: Design matrix and response values in case of the example discoloration of a clear coat

Number of experiment	Matting	In-can preservative	Thickener	Block	Value compared to the optical grey scale
1	0	0	0	A	1
2	1	0	0	A	1
3	0	0.2	0	A	4
4	1	0.2	0	A	4
5	0	0	0.8	A	1.5
6	1	0	0.8	A	1
7	0	0.2	0.8	A	2.5
8	1	0.2	0.8	A	2.5
9	0	0	0	B	1
10	1	0	0	B	1.5
11	0	0.2	0	B	4.5
12	1	0.2	0	B	5
13	0	0	0.8	B	1
14	1	0	0.8	B	1
15	0	0.2	0.8	B	3
16	1	0.2	0.8	B	2.5

In a first step, data analysis is done without considering any blocks (see Table 2.12). Thus, from each data point two replicates (y_1, y_2) are available, which have to be averaged. In addition, a standard deviation can be calculated. Due to the fact, that this standard deviation is different for each data point an average of the variances (= geometric average of the standard deviation, see Appendix 2) has to be calculated. Twice the average variance $2 \cdot s^2 = 2 \cdot 0.15$ has to be multiplied with the student factor (t(95 %, f = 14) = 2.14) and divided by the root of the number of measurements N = 16 (see Appendix 4 and Chapter 1.6). Thus, the confidence interval of the effects VB(95 %, f = 14) = 0.38) can be determined, which is essential for evaluating the significance of the effects. Afterwards the calculation of the main effects can be done by the methods described in Chapter 2.2.1 (see Table 2.12). The results show a significant positive effect regarding discoloration by the factor in-can preservative. Thus, the reason is identified and the recipe can be modified. Other components have no influence on discoloration or even reduce it.

If blocking is used, in each block every factor-level-combination is present only once. The difference between the average values of the block should not be significant. Thus, in a first step the average value is calculated for each block (Block 1: $\bar{y} = 2.19$ and Block 2: $\bar{y} = 2.44$; see Table 2.12). Afterwards the difference between each data point and this average values is calculated (see Table 2.13) and these difference values are used for further evaluation in the same way as already shown in Table 2.12. Building the difference with the average values of the blocks has no influence on the main effects, but the variance is with $s^2 = 0.09$ much lower than before with 0.125. In addition, a possible systematic difference between the two blocks was eliminated. The two average values of the blocks are slightly different. But this effect is not significant and no block effect is present. Thus, the delay due to public holiday had no influence on the data. It´s important to mention, that variance is underestimated by formation of the difference values, because also random errors are corrected. This can be taken into account by correction factors [3].

Table 2.12: Evaluation of the experiments without block variable in case of the example discoloration of a clear coat

Number of experi-ment	Mat-ting	In-can preser-vative	Thick-ener	y_1	y_2	av-er-age	s	s^2	sign (Matting)* measured value	sign(in-can preser-vative)* measured value	sign (Thick-ener)* measured value
1	-1	-1	-1	1	1	1	0	0	-1	-1	-1
2	1	-1	-1	1	1.5	1.25	0.354	0.125	1.25	-1.25	-1.25
3	-1	1	-1	4	4.5	4.25	0.354	0.125	-4.25	4.25	-4.25
4	1	1	-1	4	5	4.5	0.707	0.5	4.5	4.5	-4.5
5	-1	-1	1	1.5	1	1.25	0.354	0.125	-1.25	-1.25	1.25
6	1	-1	1	1	1	1	0	0	1	-1	1
7	-1	1	1	2.5	3	2.75	0.354	0.125	-2.75	2.75	2.75
8	1	1	1	2.5	2.5	2.5	0	0	2.5	2.5	2.5
Sum									0	0.5	-3.5
Main-effect	0	2.38	-0.88								
Aver-age value $\bar{y}$				2.19	2.44			0.125			

Table 2.13: Evaluation of the experiments with block variable in case of the example discoloration of a clear coat

Number of experiment	Matting	In-can pre-servative	Thick-ener	y_1	y_2	average	s	s^2
1	-1	-1	-1	-1.19	-1.44	-1.313	0.177	0.031
2	1	-1	-1	-1.19	-0.94	-1.063	0.177	0.031
3	-1	1	-1	1.81	2.06	1.938	0.177	0.031
4	1	1	-1	1.81	2.56	2.188	0.530	0.281
5	-1	-1	1	-0.69	-1.44	-1.063	0.530	0.281
6	1	-1	1	-1.19	-1.44	-1.313	0.177	0.031
7	-1	1	1	0.31	0.56	0.438	0.177	0.031
8	1	1	1	0.31	0.06	0.188	0.177	0.031
Main effect	0	2.38	-0.88					
Average value								0.094

This example shows, how design of experiments (DoE) can introduce more robustness to the coating system against variations in component levels and in processing factors. The question can be answered clearly and with statistical validity and is a perfect base for optimizing the formulation. Due to the block variable variance is reduced and more precise statements and predictions are possible.

However, the number of experiments is with N = 16 quite high and somebody may ask about the economics. This issue will be discussed in Chapter 2.3 and a possible solution will be presented.

2.3 *Fractional factorial designs - to separate the wheat from the chaff*

2.3.1 *Basic principle of the reduction - confounding*

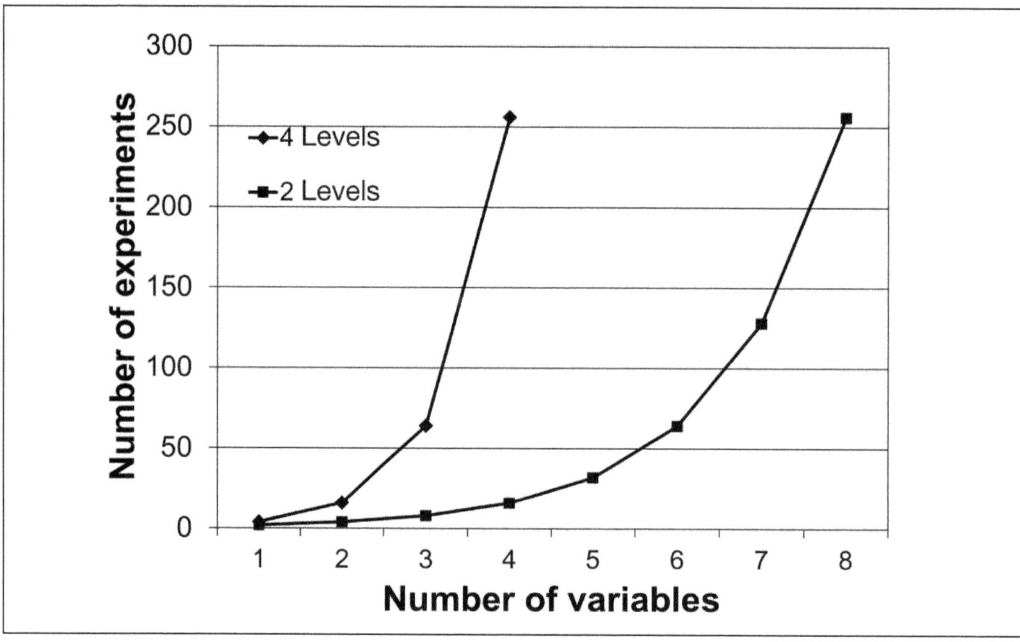

Figure 2.13: Number of experiments vs. number of variables for two levels

With an increasing number of factors the number of necessary experiments is growing very rapid. In general the number of runs can be calculated by Equation 2.12 depending on the number of variables and levels.

Equation 2.12: Number of experiments = Number of levels $^{\text{Number of variables}}$

Especially in case of a new and unknown system a lot of influencing factors are present at the very beginning (screening phase). In this step, it is necessary to identify very fast the important factors by a ranking, to minimize the experimental effort. Fractional factorial designs are a concept of DoE to accomplish an efficient experimental effort.

▶ ***Basic principle:*** Due to the concept of fractional factorial design, the number of experiments can be reduced considerably, nevertheless generating insights at the same level of quality. The aim is to separate the wheat from the chaff with a minimal number of experiments and to determine the really significant factors. The principle is based on the combination of considerable effects and not considerable side effects.

How is this decrease accomplished in the number of necessary experiments? In case of factorial designs main effects are calculated directly from the difference between the average of all

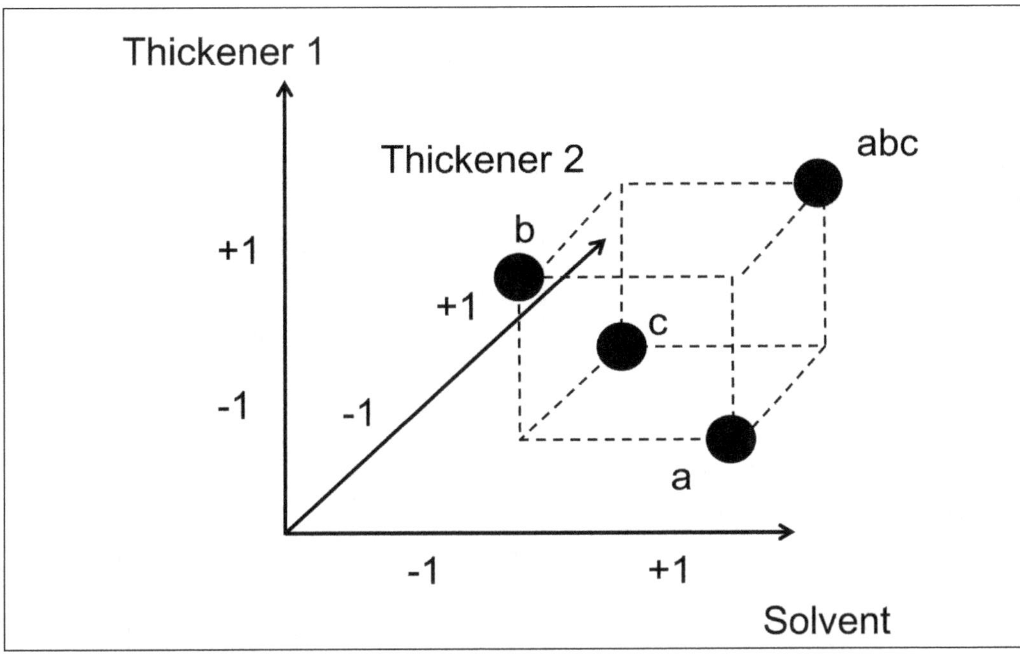

Figure 2.14: 2^{3-1}-design

target values at low and high level of the influencing factor. All other factors are at a constant level. The factors are independent from each other (principle of orthogonal designs, see Chapter 2.1.4). In contrast to full factorial designs **fractionation** takes place in case of fractional factorial designs by **mixing** (= **confounding**) some factors (e.g. one influencing factor and one interaction). Thus, a new variable is generated. Mixing indicates, that a factor and with that mixed factor (e.g. interaction) have after mixing the same sign column in the experimental design. Both factors are varied always together in the same way. Therefore, in **fractional factorial design** beside classical influencing factors (A, B, C, …) also interactions (A·B, A·B·C, …) are factors. Not mixed factors can be adjusted directly by variation of the parameter at different levels. Interactions (= multiplied combinations from single effects) and mixed variables are adjusted indirectly through this [2]. Due to this, the original parameters can no longer be computed independently. Only the sum effect of all together mixed variables on the target value can be estimated. Effects which have these properties are called **aliases**. However, although we pay this price, the number of experiments is reduced drastically. This approach is especially useful, if one of the mixed factors is negligible small (e.g. three-, four, fife-factor interactions, because it is not very likely, that an effect of one factor depends on the level of three or even more factors – see Chapter 1.5). Very often also two-factor interactions can be excluded due to deep technical knowledge of the system. Keep in mind that not all effects are of appreciable size and meaningful. In these cases mixing can be planned also with these interactions. However, very often mixing with one factor means also automatically, mixing with further factors. This is the result of the certain designs. Details are available in literature [3].

The technique behind mixing is described considering the 2^4-design. In case of a 2^3-design three factors are investigated in 8 runs. If a further factor should be analyzed, the design has to be expanded to a 2^4-design. This ends up in already 16 runs (see Table 2.9). However, at higher designs most of the columns in the evaluation table represent already interactions (see Table 2.14). Whereas the 2^3-design produces three factors, three two-factor und one three-factor interaction (see Table 2.14), the 2^4-design reveals four factors, but already six two-factor and four three-factor respective one four-factor interactions. It has been previously pointed

Table 2.14: Evaluation table of the 2^3-design

A	B	C	WW(AB)	WW(BC)	WW(AC)	WW(ABC)
-	-	-	+	+	+	-
+	-	-	-	+	-	+
-	+	-	-	-	+	+
+	+	-	+	-	-	-
-	-	+	+	-	-	+
+	-	+	-	-	+	-
-	+	+	-	+	-	-
+	+	+	+	+	+	+

Table 2.15: 2^4-design by allocating of the three-factor interaction ABC to variable 4 (mixing)

Number of experiment	A	B	C	Variable 4 = WW(ABC)
1	-	-	-	-
2	+	-	-	+
3	-	+	-	+
4	+	+	-	-
5	-	-	+	+
6	+	-	+	-
7	-	+	+	-
8	+	+	+	+
9	-	-	-	-
10	+	-	-	+
11	-	+	-	+
12	+	+	-	-
13	-	-	+	+
14	+	-	+	-
15	-	+	+	-
16	+	+	+	+

Table 2.16: 2^{4-1}-design

Number of experiment	A	B	C	Variable 4
1	-	-	-	-
2	+	-	-	+
3	-	+	-	+
4	+	+	-	-
5	-	-	+	+
6	+	-	+	-
7	-	+	+	-
8	+	+	+	+

out that especially interactions of higher order are negligible. Hence, there is a possibility to replace interactions by mixing with further factors. With the allocation of the additional fourth factor to the three-factor interaction ABC of the 2^3-design a new 2^4-design can be generated (Table 2.15). In this design some lines occur twice and can be eliminated (Table 2.15). At the end a fractional factorial 2^{4-1} design is generated (Table 2.16). The result of the evaluation of the experimental design are now "only" three main effects (factors A, B, C) and the main effect of the fourth factor combined with the three-factor interaction of the factors A, B and C. It is impossible to differentiate between variable D and the interaction ABC. Only the expansion to a factorial design would enable this.

In general such designs are called 2^{k-p} fractional factorial designs, whereas k is the number of factors and p the number of effects aliased/confounded with interactions effects.

Example: Influence of the solvent on the structural-viscosity of a waterborne, pigmented coating

Mixing can also be demonstrated by the reduction from a 2^3-design to a 2^{3-1}-design. Once more the rheological analysis of a waterborne pigmented coating from Chapter 2.2.2 is used as an example. The effect of two thickeners and the solvent on the structural viscosity is analyzed. Target value is the ration of the elastic modulus G` to the viscous modulus G`` directly out of an oscillatory-rheological measurement.

In contrast to the example in Chapter 2.2.2 a mixing of the parameter solvent and the interaction of thickener 1 with thickener 2 is done. Thus not three factors are analyzed in a factorial design. As alternative, starting from the 2^2 plan (thickener 1 and thickener 2), the interaction between these two variables is replacement by a new factor solvent. Due to this in contrast to the classical 2^3 plan with 8 experiments only 4 are necessary in case of the 2^{3-1} plan (see Figure 2.14). However, the main effect of the solvent is mixed with the interaction WW(V1V2).

If the main effect of the solvent is calculated in the **f**ractional **f**actorial design $HE(Solvent)_{FF}$ (see data in Table 2.7), the reduced form is defined by

Equation 2.13:

$$HE(Surfactant)_{FF} = \frac{abc + a}{2} - \frac{b + c}{2} = \frac{1.43 + 1.12}{2} - \frac{1.14 + 1.14}{2} = 0.135$$

The interaction of both thickener in the fractional factorial design $WW(V1V2)_{FF}$ can be calculated by

Equation 2.14:

$$HE(V1V2)_{FF} = \frac{1.43 + 1.12}{2} - \frac{1.14 + 1.14}{2} = 0.135$$

Thus, the same value appears for the main effect of the solvent and the interaction of the two thickeners. However, be aware that both values represent neither the main effect nor the interaction alone! It's only possible to calculate the mixed value of HE(Solvent) + WW(V1V2). If these values are compared with those of the factorial design in Chapter 2.2.2 and the summation of HE(Solvent) + WW(V1V2) is equal to 0.135.

2.3.2 Blocking - perfect suited for the 2^{4-1}-design

As already presented in Chapter 2.1.2 it is sometimes useful to split up the data set in blocks to prevent distortion due to perhaps existing trends or other systematic differences. For example, if two different batches of a raw material are necessary or one part of the experiments is performed in week 1 and the other in week 2 blocking might be helpful. In such situations the fractional factorial 2^{4-1} design (see Table 2.16) perfectly suits, because the three-factor interaction can be used as block variable (see Table 2.17). Due to this it is possible to study the influence of the block variable and to reduce the number of experiments.

Table 2.17: 2^{4-1}-design for blocking

Number of experiment	A	B	C	Variable 4 = blockvariable
1	-	-	-	Block 1
2	+	-	-	Block 2
3	-	+	-	Block 2
4	+	+	-	Block 1
5	-	-	+	Block 2
6	+	-	+	Block 1
7	-	+	+	Block 1
8	+	+	+	Block 2

Example: Stability of a biocide in a coating

Subject of an experimental series is the improvement of the biocide stability in a coating. Target value is the recovery rate (RR) of the biocide after 8 weeks of storage at 40 °C. Influencing factors are the ratio of stabilizer/biocide, dryer A

Table 2.18: Experimental parameters and the experimental area for the example recovery rate of a biocide

Factors	Low limit in [%]	High limit in [%]
Stabilizer/ biocide-ratio	0.4	1.0
Biocide batch	A	B
Dryer A	0	0.6
Dryer B	0	0.4

Table 2.19: Design matrix for the example recovery rate of a biocide

Name of experiment	Number of experiment	Stabilizer/ biocide-ratio	Dryer A	Dryer B	Biocide batch	RR
1	8	-1	-1	-1	-1	78
2	5	1	-1	1	-1	87
3	4	-1	-1	1	1	87
4	6	1	-1	-1	1	99
5	2	-1	1	1	-1	79
6	7	1	1	-1	-1	88
7	1	-1	1	-1	1	88
8	3	1	1	1	1	97

and dryer B. A full factorial 2^3-design might be the best solution. However, due to the small and limited amount of raw materials it is necessary to distinguish between two different batches of the biocide. Thus, in this example also the difference between the two batches on the recovery rate RR is examined. A 2^4-design will be the result, if the block variable (batch 1 and batch 2) is only used as additional variable (see Chapter 2.2.5). The question can be analyzed very efficiently with a fractional factorial design, if the usually not significant three-factor interaction is confounded with the block variable. Thus we end up with a 2^{4-1}-design.

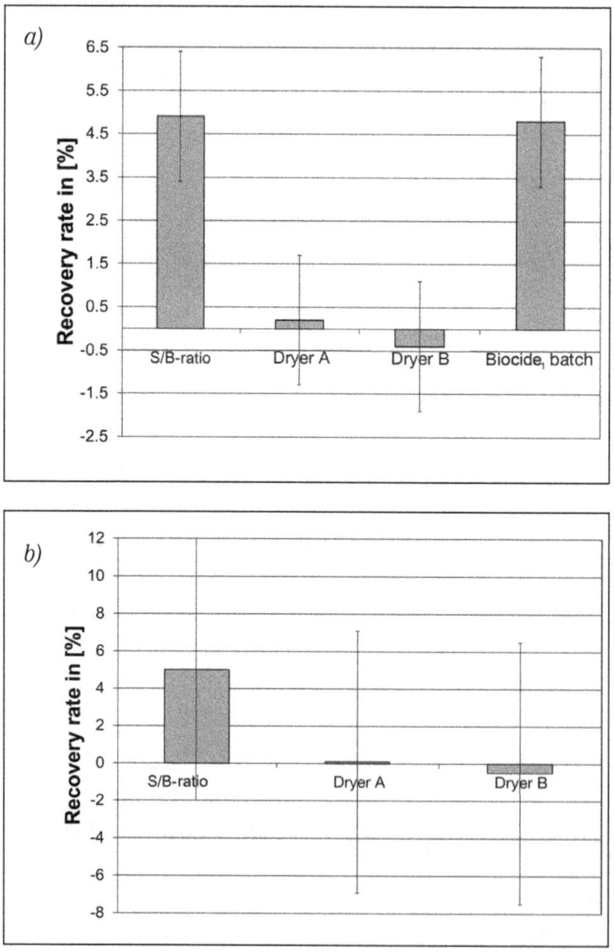

Figure 2.15: Diagram of the model parameters (=effects/2) for the example recovery rate of a biocide.
a) taking blocking into consideration, b) without blocking

Properties of the experimental design (see Table 2.19):

- Two-level factorial design
- Number of factors: 4 (see Table 2.18)
- Number of blocks: 2
- Number of target values: 1
- Number of experiments: 8
- Randomization: yes

Only the ratio of stabilizer to biocide is relevant for the stability of the biocide in the coating (see Figure 2.15a and 2.16). The amount of driers is no critical formulation parameter regarding the recovery rate. However, it is very important to realize, that there is huge significant difference between both blocks. Thus, the two batches differ in their influence on the recovery rate.

2.4.1 Central composite designs

Central composite designs are factorial designs with centre points and additional so-called **star runs**, symmetrically arrayed on the factor axes (see Figure 2.19 and 2.20). In addition to the typical labelling -1, +1 and 0 the distance +/- α of the axial runs from the design centre must be specified (see Figure 2.21, Table 2.23). Also in this design the centre point is perfectly suited for replicate runs. Core of the design is a full factorial design or also a fractional factorial 2^{k-p}-design from type V. The number of experiments for two levels is defined by Equation 2.15.

Equation 2.15: $N = 1 + 2 \cdot k + 2^{k-p}$

k = number of factors
p = number of effects confounded with other effects

Table 2.23: Experimental design of a central composite design with star runs

Number of experiment	A	B	
1	-	-	Edge points
2	+	+	
3	-	+	
4	+	-	
5	0	0	Centre point
6	$-\alpha$	0	Star runs
7	$+\alpha$	0	
8	0	$-\alpha$	
9	0	$+\alpha$	

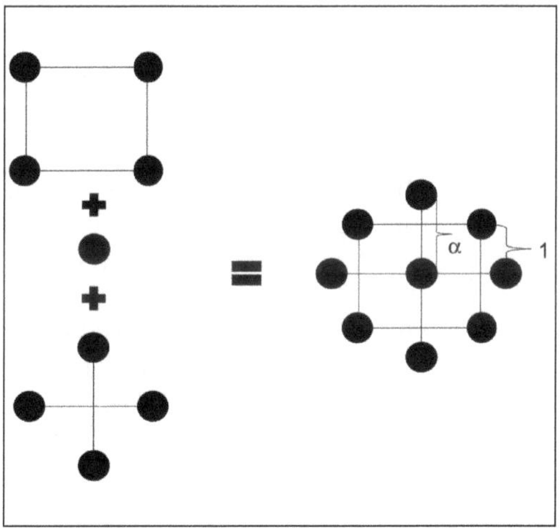

Figure 2.21: Central composite design for two factors

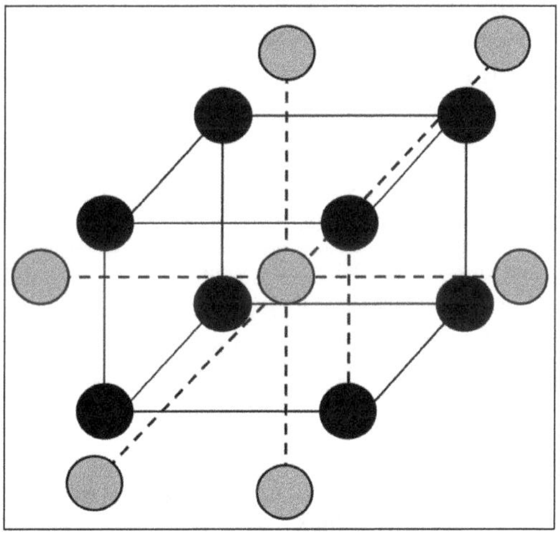

Figure 2.22: Central composite design for three factors

Sometimes, the experimental setup cannot be adjusted exactly to the values $-\alpha$, -1, 0, +1, $+\alpha$, +a. For example it is only possible to adjust the pressure in steps of 0.5 bar, because the controller is not continuously variable. If a value of 0.3 bar or 1.7 bar is necessary, it is impossible to adjust the pressure in the experimental setup. In such situations, α has to be adjusted exactly to the real dataset. In addition in case of star runs it is still necessary that they can be realized. Keep in mind, star runs are outside the experimental area! Especially in case of concentrations star runs are always a problem, if the level should be below 0 %.

The following central composited designs are very often used in industrial practice:

- **CCD-Plan – central-composite-design** with $\alpha = 1$ and a 2^2-design up to $\alpha = 1.761$ in case of 2^6-designs; no mixing is present, very small confidence intervals; very useful in case of a small number of experimental runs and if no confounding should take place.

- **CCF-Plan – central-composite-faced (or RSM design)**, meaning $\alpha = 1$; star runs are on the surface of the cube; the design is not orthogonal for all effects and interactions and should be used only in exceptional cases. This design is useful, if it doesn´t make sense to operate outside the experimental area.
- Regarding the **rotatable design** and the **pseudo orthogonal, rotatable design** we refer to the special literature [3].

Example: Minimization of the coating thickness in case of a base coat

The aim is to minimize the dry film thickness of a silver-metallic effect base coat and to keep the colour shade as well as all other technical specification at constant level. At present the dry film thickness is 10 μm. Starting from a recipe the spread rate should be optimized due to variation of the recipe parameters 1 and 2. Response is the dry film thickness, with which a perfect hiding power of the substrate is possible. Thus, films are applied in different thicknesses to charts that offer black and white areas. After drying, the surface is examined by colour measurement. The layer thickness (in micrometres) at which a contrast between black and white areas is not detectable is determined and is taken as a response of the hiding power (following EN ISO 13300).

Properties of the 2^2-CCD-design with central point and star runs ($\alpha = 1.412$) (see Table 2.25):

- Number of factors: 2 (see Table 2.24)
- Number of blocks: 1
- Number of target values: 1
- Number of experiments: 10, including 2 replicates in the centre point
- Randomization: yes

Table 2.24: Factor settings regarding the example optimization of the coating thickness in case of a base coat

Factors	Low limit in [%]	High limit in [%]
Parameter 1	28	35
Parameter 2	15	20

Table 2.25: Design matrix regarding the example optimization of the coating thickness in case of a base coat

Number	Factors		Response
	Parameter 1	Parameter 2	Hiding power in [μm]
1	31.5	17.5	6
2	29	16	7
3	34	16	9
4	29	19	8
5	34	19	6
6	27.97	17.5	8
7	35.04	17.5	9
8	31.5	15.38	6
9	31.5	19.62	5
10	31.5	17.5	6

The system is due to a strong interaction non-linear (see Figure 2.23 and 2.24) and a saddle shaped surface is resulting (see Figure 2.25). From the response surface diagram or the contour plot a minimal dry film thickness of 5 μm can be derived (Figure 2.25). The correlating composition at the minimum is regarding parameter 1 at 32.6% and parameter 2 at 19.6%. The optima would be a very bad place to set up your process, because results would be very sensitive to variations in the input factors. However, keep in mind the starting point of 10 μm. Due to this insight about the non-linear behaviour the coating can be used much more efficient and also sustainable. Even if in practice only a reduction to 8 μm is realized, still an improvement of 20 % is possible. Without the systematic approach of DoE this would not be possible.

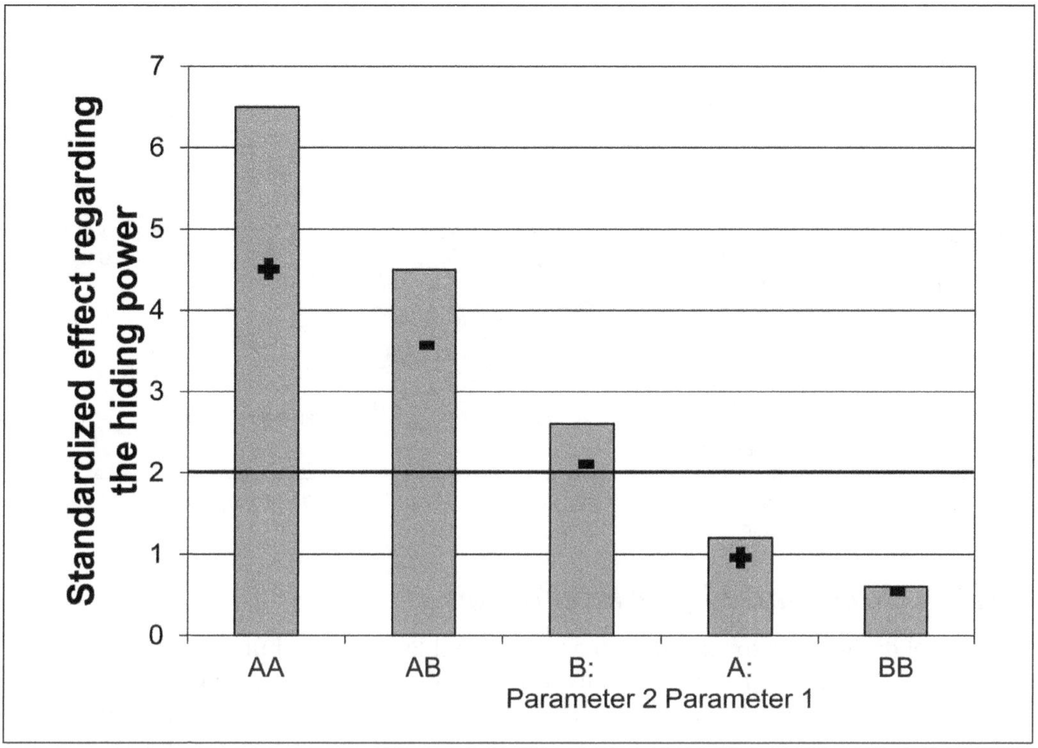

Figure 2.23: Pareto-Diagram regarding the example optimization of the coating thickness in case of a base coat

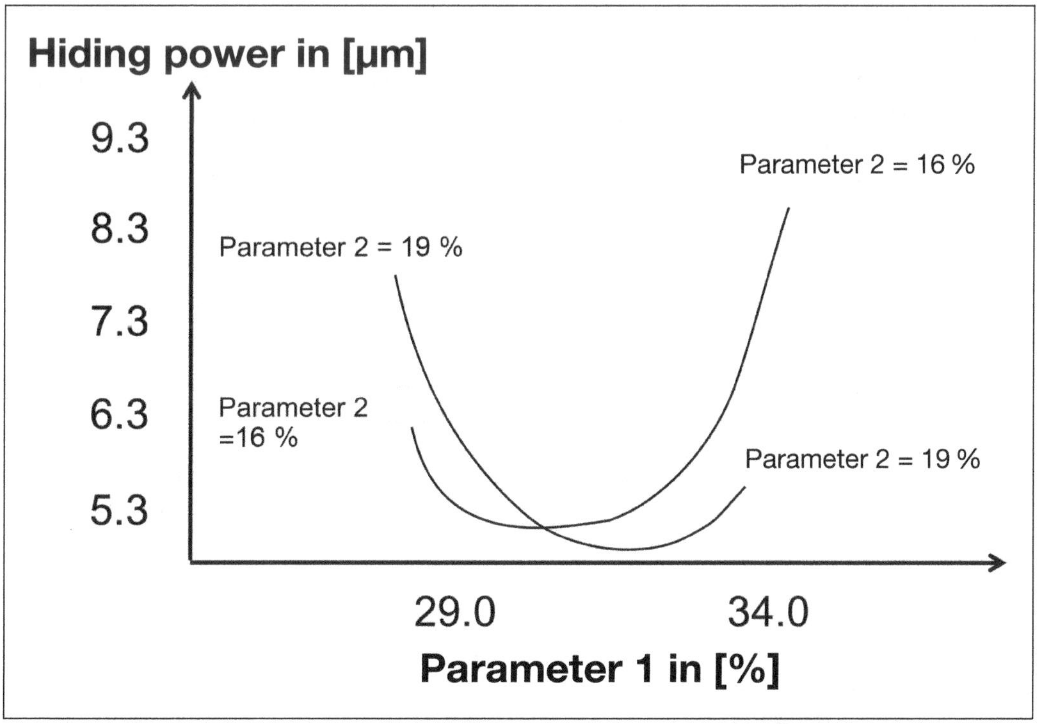

Figure 2.24: Interaction diagram regarding the example optimization of the coating thickness in case of a base coat

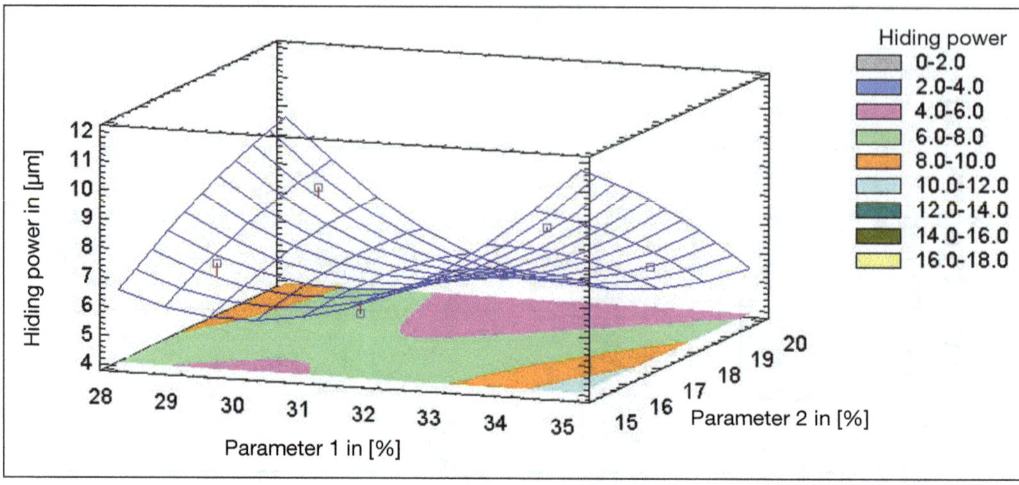

Figure 2.25: Response surface incl. contour plot regarding the optimization of the coating thickness in case of a base coat. Response is always the hiding power in μm.

2.4.2 Three- and higher level designs

If two-level designs are extended with an additional level, we end up with k factors automatically at 3^k-designs. The number of experiments is increasing even much faster than in case of 2 levels, but also in this case fractional factorial design like 3^{k-p} are possible. In such designs a confounding of main effects and interactions take place, which is only acceptable for screening designs. In case of modelling it is not recommended to determine the quadratic terms of the model without any independent interactions.

2.4.3 Mixed designs

The combination of standard designs to so-called mixed designs is also possible. Partially more experiments are necessary than in case of full factorial plans (see Chapter 2.2) or also central composite designs (see Chapter 2.4.1 and Table 3.26). Combining factors with two- and three-levels the core of the design is still a two-factorial design. At each edge of the core the three-level factor is varied over all levels. Reducing the degree of levels diminishes also the number of experiments. Unfortunately the ability to evaluate non-linear effects is lost. After harmonization of the levels to one degree (e.g. two-level), it is also possible to analyze the factors in a central composite design. If some factors are only adjustable at two levels (i.e. with qualitative influencing factors), it is also possible to perform a central composite design at the two edge points +1 and -1 with those factors, that have more than two levels.

Table 2.26: Standard design in case of 5 factors with different distribution between two and three levels [2]

Number of two-level factors n	Number of three-level factors m	Number of experiments in case of a full factorial design	Number of experiments in case of a fractional factorial core
1	4	$2^1 * 3^4 = 2 * 81 = 162$	$2^{1-1} * 3^4 = 81$
2	3	$2^2 * 3^3 = 4 * 27 = 108$	$2^{2-1} * 3^3 = 54$
3	2	$2^3 * 3^2 = 8 * 9 = 72$	$2^{3-1} * 3^2 = 36$
4	1	$2^4 * 3^1 = 16 * 3 = 48$	$2^{4-1} * 3^1 = 24$
5 (with a center point and star runs)	-	25 + 10 + 1 = 43	$2^{5-1} + 10 + 1 = 27$
5	-	$2^5 = 32$	$2^{5-1} = 16$

Example: Optimal levelling of a clear coating

Table 2.27: Factor settings for the example levelling of a clear coating

Factors	Low limit in [%]	High limit in [%]	levels
Binder	0	10.0	3
Additive	0	0.5	3
Solvent 1	0	3.0	3
Solvent 2	0	4.0	2

To optimize the balance between levelling and sagging of a clear coat, 4 coating components are analyzed in a full-factorial design. In preliminary studies these 4 components has been identified as most important effects. Levelling is analyzed at a film thickness of 30 μm with the wave-scan method regarding the distinctness of image DOI (0 to 100, with 100 the best value). This value is calculated from the response dullness (du) and the degree of long and short waves (Wa, Wb). Aim is to maximize the uniform appearance of the coating [8].

Properties of the mixed multi-level design (see Table 2.28):

- Number of factors: 4 (see Table 2.27)
- Number of blocks: 1
- Number of target values: 1
- Number of experiments: 54
- Randomization: no

Table 2.28: Design matrix regarding levelling of a clear coating. In the formulation only the components with the most important effects are sometimes at level 0. Beside these factors also other components are present, which are still sufficient to formulate a coating.

Number	Factors				Response
	Binder	Additive	Solvent 1	Solvent 2	DOI 30 μm
1	0	0	0	0	82.9
2	5	0	0	0	84.5
3	10	0	0	0	82.6
4	0	0.25	0	0	83.8
5	5	0.25	0	0	83.3
6	10	0.25	0	0	83.9
7	0	0.5	0	0	82.1
8	5	0.5	0	0	84.9
9	10	0.5	0	0	84.5
10	0	0	1.5	0	83.9
11	5	0	1.5	0	82.7
12	10	0	1.5	0	82.1
13	0	0.25	1.5	0	85.4
14	5	0.25	1.5	0	86.6
15	10	0.25	1.5	0	85.7
16	0	0.5	1.5	0	85.6
17	5	0.5	1.5	0	85.4
18	10	0.5	1.5	0	86
19	0	0	3	0	84.6

Table 2.28 (Continue)

Number	Factors				Response
	Binder	Additive	Solvent 1	Solvent 2	DOI 30 µm
20	5	0	3	0	84
21	10	0	3	0	83.2
22	0	0.25	3	0	84.2
23	5	0.25	3	0	83.9
24	10	0.25	3	0	82.1
25	0	0.5	3	0	85.7
26	5	0.5	3	0	85.2
27	10	0.5	3	0	84.9
28	0	0	0	4	85.9
29	5	0	0	4	86.1
30	10	0	0	4	82.1
31	0	0.25	0	4	84.8
32	5	0.25	0	4	84.5
33	10	0.25	0	4	82.5
34	0	0.5	0	4	84.2
35	5	0.5	0	4	83.3
36	10	0.5	0	4	84.1
37	0	0	1.5	4	84.2
38	5	0	1.5	4	83.7
39	10	0	1.5	4	80.5
40	0	0.25	1.5	4	84.3
41	5	0.25	1.5	4	84.9
42	10	0.25	1.5	4	84.6
43	0	0.5	1.5	4	85.3
44	5	0.5	1.5	4	85
45	10	0.5	1.5	4	83.4
46	0	0	3	4	84.3
47	5	0	3	4	85
48	10	0	3	4	81.8
49	0	0.25	3	4	85.2
50	5	0.25	3	4	84.3
51	10	0.25	3	4	84
52	0	0.5	3	4	84.8
53	5	0.5	3	4	85.2
54	10	0.5	3	4	86.3

The additive offers the most important effect on DOI and also the binder has a significant effect (see Figure 2.26). It is not possible to neglect the interaction between these two factors (see Figure 2.27). The effects of both solvents are not significant and below the 2s-limit (see Figure 2.26). In Figure 2.28 a response surface model with a strong non-linear behaviour is shown for the experimental area. A maximum in DOI become apparent at a high amount

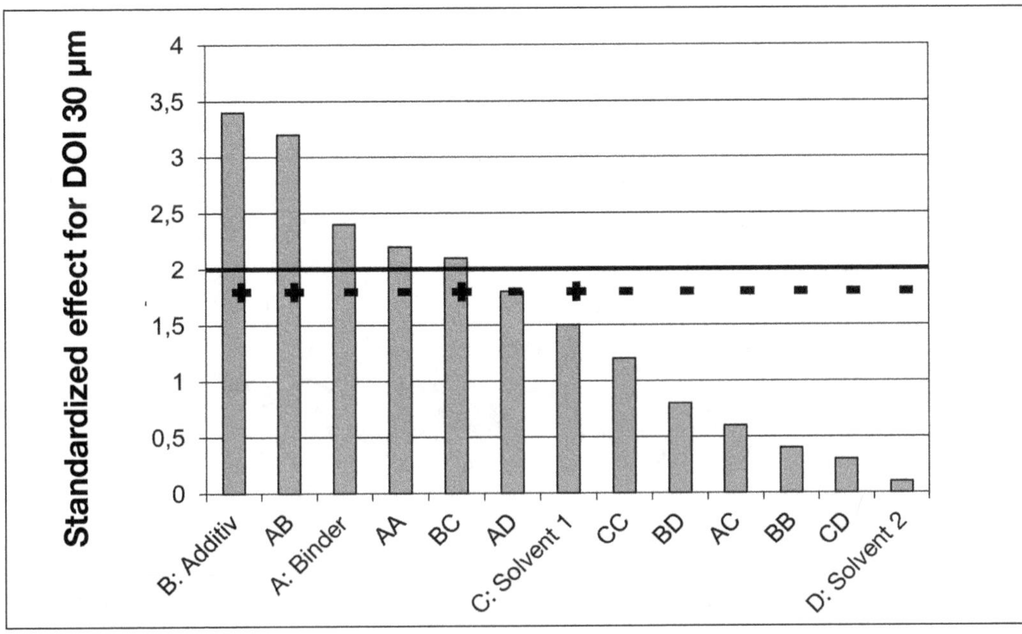

Figure 2.26: Pareto-diagram for the example levelling of a clear coating

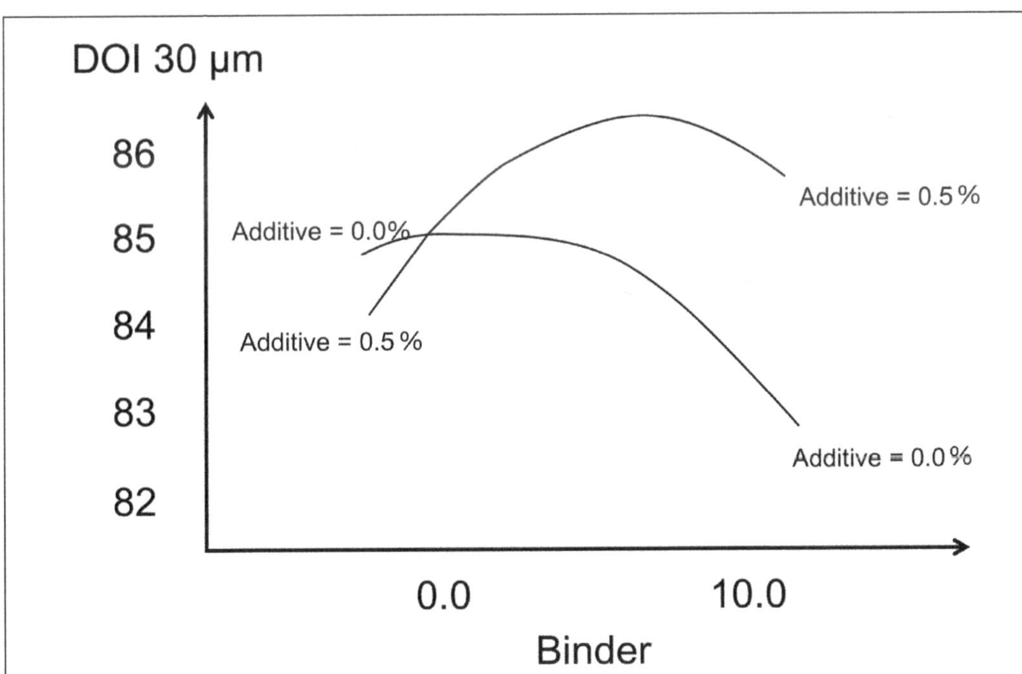

Figure 2.27: Interaction diagram for the example levelling of a clear coating

of additive and medium size binder content. This example illustrates how DoE supports the generation of system knowledge. A good technician knows very much about "his" coatings. However, this knowledge should be described as experience, because it's hardly any quantitative. A feeling is available, but such an analysis complements this experience with knowledge.

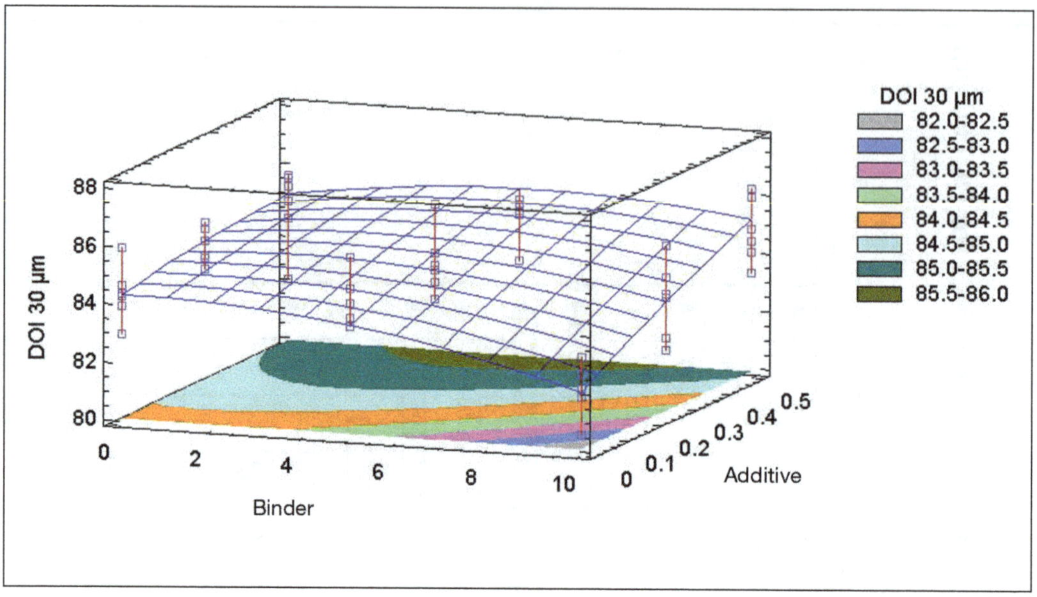

Figure 2.28: Response surface incl. contour plot for the example levelling of a clear coating; solvent 1: 3 %, solvent 2: 2 %

2.4.4 Box-Behnken designs

Box-Behnken designs have a similar design as 3^{k-p}-designs in case of three levels [4]. In this case also two-factor interactions can be analyzed. In case of three factors the design is always of a full factorial design for each possible combination of the factors (AB, AC, BC, etc.) and a centre-point (Table 2.29). One of the factors is always fixed at level 0 (Table 2.29) and we leave out the edges of the cube. Thus, the total number of experiments is smaller than in case of a CCD-design, but quadratic models cannot be examined exactly.

Table 2.29: Box-Behnken-design for three factors [3]

Number of experiment	A	B	C	Remarks
1	-1	-1	0	Full factorial design for factors A, B
2	1	-1	0	
3	-1	1	0	
4	1	1	0	
5	-1	0	-1	Full factorial design for factors A, C
6	1	0	-1	
7	-1	0	1	
8	1	0	1	
9	0	-1	-1	Full factorial design for factors B, C
10	0	1	-1	
11	0	-1	1	
12	0	1	1	
13	0	0	0	Centre point
14	0	0	0	
15	0	0	0	

2.4.5 D-optimal designs - egg-laying wool-milk-sow

In practice very often situations appear, in which all until now described standard designs are not perfectly suited. If many factors are present, very often also fractional factorial designs result in too much experiments. This is especially the case, if someone would like to use a design with resolution IV (see Chapter 2.3.3), because this resolution is optimized for screening designs. The so-called D-optimization supports to reduce the number of experiments even below the level of fractional factorial designs. Whereas in case of 9 factors and a full factorial 2^9-design 512 experiments are necessary, it is possible to reduce the number to 32 with a fractional factorial design of resolution IV. However, in case of D-optimal designs the minimal number of experiments is twice the number of factors. Thus only 18 experiments are necessary (example in Chapter 2.6) and a considerable reduction is possible. Mostly, but not always, the statistical validity is diminished. D-optimal designs offer the possibility to choose the experimental area and fit it to a model. (see Chapter 2.1.5). This is very important, because usually the model has to be adjusted to the design. Thus, these designs are very often used in case of experimental areas which have not the shape of standard designs. Even parts of a model can be chosen and selectively adjusted to reality. This is an advantage, if a lot of knowledge is available about a system. Certain not possible combinations can be excluded already during the planning. Already available results, which does not suit perfect to the design can be taken into account. Usually central composited designs are the favourite, but in such situations D-optimal plans might be the best solution. These designs are not orthogonal and calculated by algorithms with software packages. Beside D-, also A-, E-, G- and I-optimal designs are available. However, still D-optimal versions are the most important. For more information please consider the literature [3].

2.5 Mixture design - a huge area

In mixture designs the response is a function of the component ratio only and not of its total quantity [5, 6]. It is evident that the components level-ratios are not independent and that the level of one component depends on levels of all other components. Mixtures are omnipresent in coatings technology. Similar applications can be found in food industry (e.g. ice, cake mixtures) or with building material (e.g. concrete, glass). For sure not always a mixture design is used in these fields, because very often the analyzed factors are only present in a small amount (e.g. additive) and all other components are predominant. Then full or fractional factorial designs are better suited. Two tricks help to avoid mixture designs:

- The mixing ratio is used as factor itself in the design (ratio of binder to solvent), or
- one component is always at a constant level, because it is in a predominate surplus

Table 2.30: Mixture design with three components

X_1	X_2	X_3	Response Y
1	0	0	Y_1
0	1	0	Y_2
0	0	1	Y_3

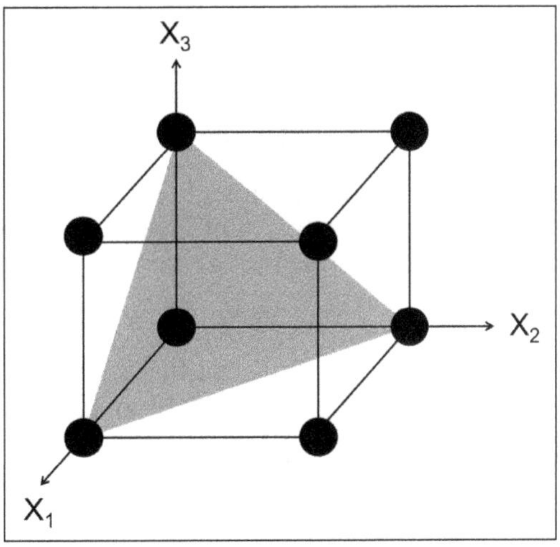

Figure 2.29: Mixture design for three components

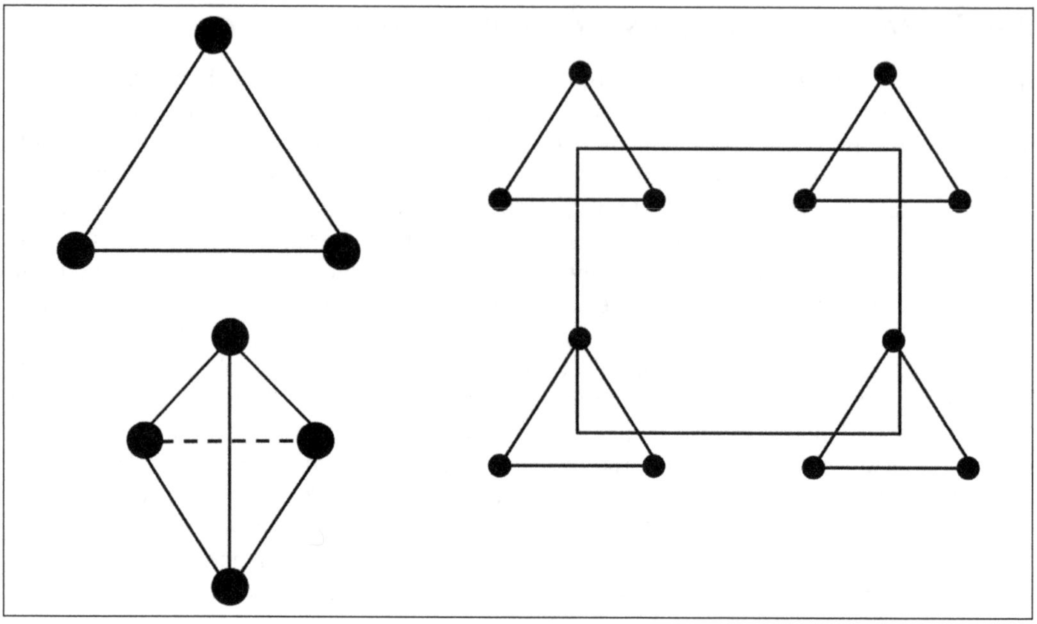

Figure 2.30: Higher mixture designs with a simplex lattice design: Triangular (3 components), Tetrahedron (4 components), combination of a 3-component mixture design (triangular) with a 2^2-design due to the position of the triangular at each edge of the 2^2-design (5 components).

In case of mixture designs every factor can vary between 0 and 1 and negative quantities must be avoided. The situation is mathematically expressed in Equation 2.16 with three mixture factors ranging between 0 and 1.

Equation 2.16: $1 = x_1 + x_2 + x_3$

Thus, the experimental area is reduced to a triangular, which is equal to the well known **three-component diagrams** of *Rozebum* (see Figure 2.29). An introduction using trinary diagrams is given in Appendix 6. These arrays are no longer orthogonal, but still the same kind of diagnostic and modelling tools can be used as already described.

▶ ***Basic principle:*** The characteristic feature of a mixture is that all ingredients sum to 100 %. Hence, the response is only a function of the component ratio and not of its total quantity. In such situations mixture designs should be applied.

More than three components are illustrated in tetrahedrons or in case of even more factors in so-called **simplex-lattice designs** [5, 6] (see Figure 2.30).

Also in case of mixture designs a model can be determined to predict the behaviour for each composition. The simplest mixture design examines only main effects. Thus, a linear model can be derived applying Equation 2.16 (see Equation 2.17).

Equation 2.17: $y(x_1, x_2, x_3) = a_0 + a_1x_1 + a_2x_2 + a_3x_3$

The determination of the coefficients is simple, because they are equal to the target values at each data point (Equation 2.18 to 2.20).

Equation 2.18: $a_1 = y(x_1 = 1, 0, 0) - a_0$

Equation 2.19: $a_2 = y(0, x_2 = 1, 0) - a_0$

Equation 2.20: $a_3 = y(0, 0, x_3 = 1) - a_0$

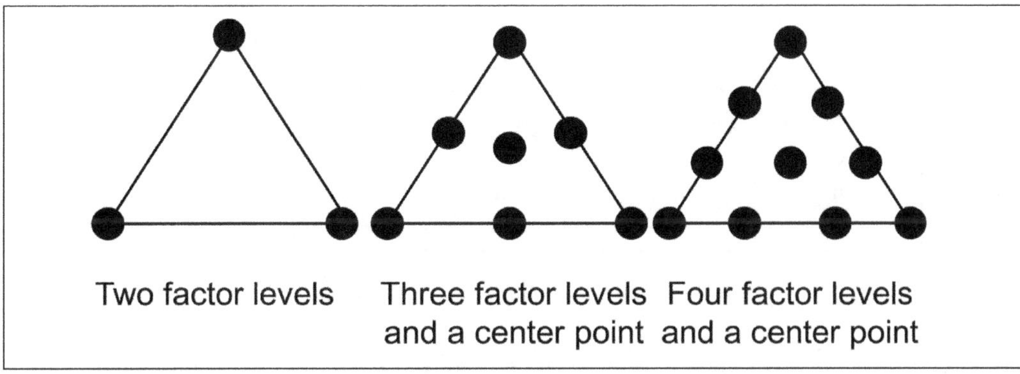

Figure 2.31: Higher levels in mixture designs with a simplex lattice design for three components

Naturally data points can be removed and fractional designs generated. It is also possible to add data points, which offers the possibility to examine interactions. However, the number of experiments is increasing drastically (Figure 2.31 and Table 2.31).

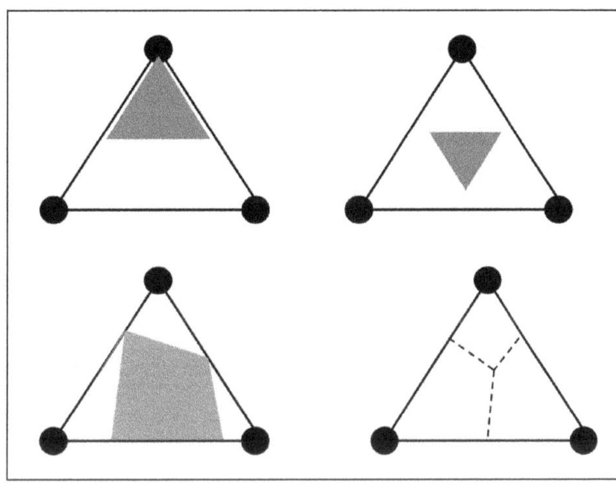

Figure 2.32: Not-realizable, irregular areas (grey) and border lines due to phase changes (dotted line) in mixture designs

Especially in case of phase equilibriums it is very often the case that some physical limits are present in the system and certain component ratios are impossible (maximal amount of solvent in a coating, dissolution border, etc.) (grey shaded area in Figure 2.32). Simplex designs are perfectly suited for such situations. Otherwise in case of very complex systems also D-optimal designs might be used. If there are some borders in the system, two separate designs has to be generated (see Figure 2.32). Even the combination of full factorial designs with mixture designs is possible.

Example: Adhesion optimization of a base coat

A new medium-solid base coat with good adhesion should be developed based on the mixing of three different binders. The total amount of the binder in the coating was fixed by 50 %. Target value is the adhesion SWK GT regarding cross-cut test based on EN ISO 2409 (a 5 step classification from perfect adhesion 0 to no adhesion 4 is used) after exposing the test specimens to condensation-water atmospheres (following EN ISO 6270-2).

Table 2.31: Mixture design with centre-point and three levels at the axis - this is equal to the middle illustration in Figure 2.31

X_1	X_2	X_3	Y
1	0	0	Y_1
0	1	0	Y_2
0	0	1	Y_3
1/2	1/2	0	Y_4
0	1/2	1/2	Y_5
1/2	0	1/2	Y_6
1/3	1/3	1/3	Y_7

Table 2.32: Factor settings for the example adhesion of a base coat

Factors	Low limit or bound in [%]	High limit or bound in [%]	Number of levels
Binder 1	0	50.0	
Binder 2	0	50.0	4
Binder 3	0	50.0	

Table 2.33: Design matrix for the example adhesion of a base coat

Number	Factors			Target
	Binder 1	Binder 2	Binder 3	SWK GT
1	50	0	0	0.5
2	33.3333	16.6667	0	0.5
3	33.3333	0	16.6667	0.5
4	16.6667	33.3333	0	0.5
5	16.6667	16.6667	16.6667	4
6	16.6667	0	33.3333	0.5
7	0	50	0	0.5
8	0	33.3333	16.6667	0.5
9	0	16.6667	33.3333	0.5
10	0	0	50	2.5

The aim is a perfect adhesion after exposing the test specimens to condensation-water atmospheres. The target value should be minimized with a value < 0.5.

Properties of the mixture design (see Table 2.33):

- Simplex lattice
- Number of components: 3 (see Table 2.32)
- Number of target values: 1
- Number of experiments: 10
- Randomization: no

The optima regarding adhesion can be achieved by mixing of binder 1 and 3 with only a small amount of binder 2. Otherwise it's also possible to get good results mixing binder 2 and 3 and only a small amount of binder 1 (see Figure 2.33). A combination of all three binders is counterproductive regarding adhesion.

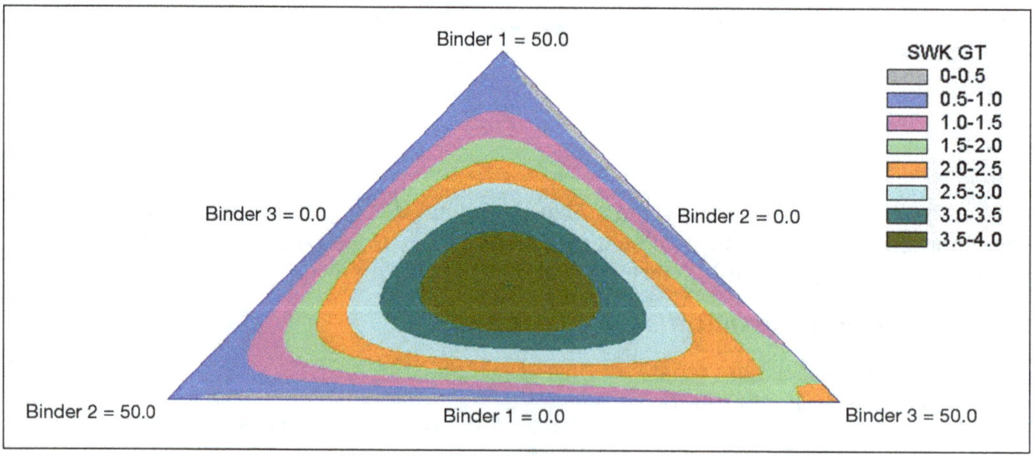

Figure 2.33: Response surface as contour plot for the effect of coating components on the adhesion of a base coat

2.6 *Qualitative classification*

Until now in this book target values and factors were always of quantitative nature. However, in surface science exists a strong correlation with feelings, scratching and optical demands. Thus, very often only qualitative based data are available (i.e. weathering tests, chemical resistance, etc.). If **target values** are of **qualitative nature**, very often school grades are used to generate at least semi-quantitaive data. Blocking or adhesion is evaluated with such a scale of grades also in standards. Such grades do not follow a normal dis-

3.1.2 Outliers

Deviating experiments that appears to differ excessively from the average (so called **outliers**) may degrade the predictive ability regarding effects and interactions and blur the interpretation of models. There are several reasons:

- The data may have been entered wrongly.
- The experiment may have been incorrectly carried out.
- The model is inappropriate in the locality of the outlier. For example the system may change fundamentally as we move into or close to irregular regions (Chapter 2.1.5).

Several detection tools are known to decide whether to retain or reject a result (e.g. N-probability plot, residual plots as well as the replicate plot - see Chapter 3.4). Please find more details in general statistics literature [3, 4]. If the deviating point is uninteresting, then remove it from the analysis. However, it is always good to repeat the experiment and sharpen the judgement. Unfortunately no universal rule can be invoked about the presence of outliers. The limits depend on the data set itself. We have to avoid retaining results that have a correct effect on the average of the data just by too stringent standards.

3.2 Confidence intervals - where are the limits?

As already considered in Chapter 1.6 and Appendix 4 and 5, it is necessary to define a range within we may reasonably assume that certain parameters describing the data set have their true value. These ranges, based on data set fluctuations has to be taken into account in all analysis steps. Thus, main effects and interactions from DoE require always a range, the **confidence interval**. In the *Pareto*-plot this is usually obvious by the 2s-border line (derived from the geometric mean of the standard deviations from all plotted main effects and interactions) or by error bars of each single effect (e.g. see Figure 2.15). Instead of the absolute scale it is common to use the standardized effect based on a division of the effect by its standard deviation. The border line of the confidence interval is in this scale for 95 % probability at 2 (e.g. see Figure 2.17 and 2.23).

The evaluation of significance of effects and interactions just by the confidence interval is only valid in case of two-level factors. Only in this case the effect can be derived directly from the difference between two average values (= comparison of the means of two samples). In case of higher level designs regression analysis has to be applied (see Chapter 3.3).

3.3 Regression - the best model

3.3.1 Basic principles

Modelling has been already described in Chapter 1.7. It has also been emphasized how main effects and interactions can be derived directly from the linear model (see Figure 1.27 and 1.28). In case of higher level designs this procedure cannot be applied, because the effects cannot be derived by simple calculation of the difference. In this case, the optimal model is calculated by the so-called regression analysis and real effects are estimated by the model coefficients.

The common procedure is described now briefly. The investigator must try to derive a "best" straight line from the points, which is based on the intuitive seeking the line that minimizes the sum of deviations between the data points and the line. The so called **regression analysis** based on the **method of least squares** provides the means for objective obtaining such a line, which must not be necessarily a straight line [1, 2, 7, 8]. First of all we define a

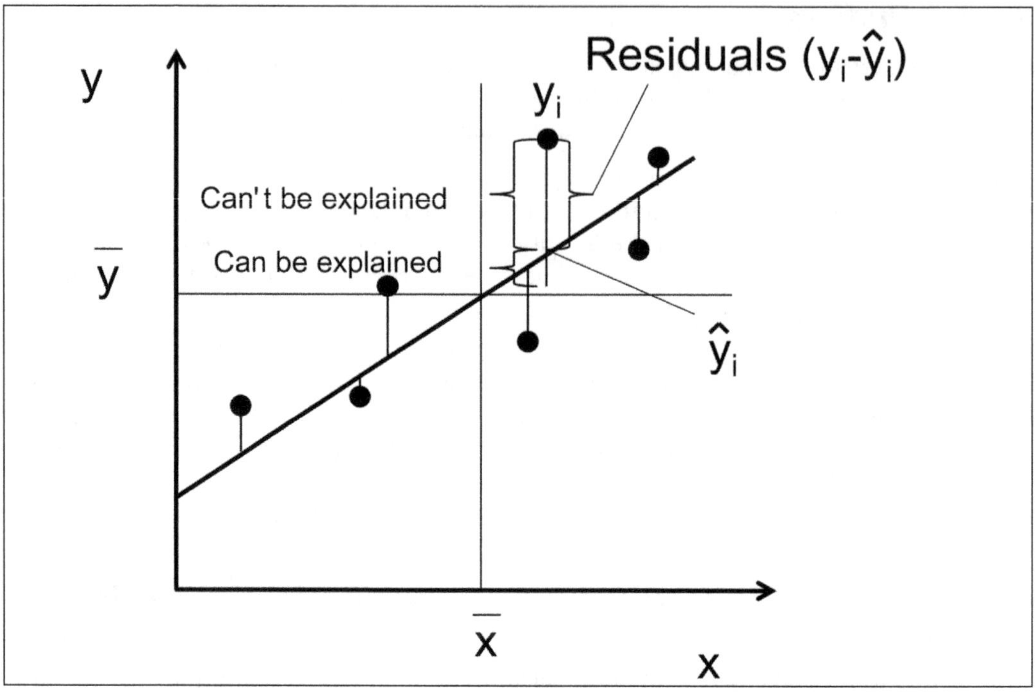

Figure 3.1: Method of the least-squares

certain function $\hat{y}$ (e.g. a straight line) and calculate the points on the calculated regression line. A good fit minimize the sum of deviations between the experimental points y_i and the calculated line $\hat{y}_i$. Since some of these deviations (commonly also known as **residuals**) will be positive and some others negative the residuals will be squared. Afterwards it is possible to seek for the minimum of the sum of squares (mathematicians write Σ) of residuals. Thus, the straight line required is calculated on this principle by seeking the function which minimizes the sum of **squares of the residuals** (SQ), which is based on solving a linear equation system (see Figure 3.1):

Equation 3.1: $SQ_{Residuals} = \Sigma_{i=1}^{N} (y_i - \hat{y}_i)^2$ => Minima

The procedure is described more in detail in literature [1, 7] and in Appendix 7 and is offered in many software packages. Also "EXCEL" offers a simple tool to calculate so called trend lines. Please find also an example in Chapter 3.3.4. However, it is important to understand the sense of the resulting parameters.

▶ ***Basic principle:*** Models are fitted to a set of data by regression analysis and the method of least-squares by minimization of the residuals. Residuals are the difference between the data point and the calculated model. Real effects are estimated by the model coefficients.

The simplest model from a perspective of mathematics is the linear model with the algebraic form:

Equation 1.10: $y = a_0 + a_1 \cdot x$

y = response
x = factor
a_1 = slope of the line
a_0 = axis intercept

The result of the slope and intercept calculations are depicted in Figure 3.1. The line at the centroid point $\bar{y}$ indicates the grand average of all target values. If we look more in detail on the certain data point y_i slightly above the grand average value $\bar{y}$, it is obvious that the corresponding x value is also above the grand average $\bar{x}$(there is an increasing trend). However, regression analysis gives only a prediction regarding the increase of $\hat{y}_i$.Thus, there is only a partly explanation for the deviation between y_i and $\bar{y}$ by the model. From the total deviation $y_i - \bar{y}$ the amount $y_i - \hat{y}_i$ cannot be explained. v

In order to estimate how well the experimental points fit to the straight line of the model given by Equation 1.10, we define the quantity $\mathbf{SQ_{Total}}$ (= **total sum of squares, TSS**), the sum of squares of the deviations from the mean $\bar{y}$ for the individual y values (Equation 3.2).

Equation 3.2: $SQ_{Total} = \Sigma_{i=1}^{N} (y_i - \bar{y})^2$ mit $\bar{y} = 1/N\ \Sigma_{i=1}^{N} y_i$, the **grand average value of the whole dataset**

Partitioning this quantity, Equation 3.2 can be written as:

Equation 3.3: $SQ_{Total} = \Sigma_{i=1}^{N} (y_i - \bar{y})^2 = \Sigma_{i=1}^{N} (\hat{y}_i - \bar{y})^2 + \Sigma_{i=1}^{N} (y_i - \hat{y}_i)^2,$

The second part is already known from Equation 3.1 as residuals. Thus, we can derive by defining a new square sum $\mathbf{SQ_{Regression}}$, the regression total sum of squares, which describe the deviation by the model:

Equation 3.4: $SQ_{Total} = SQ_{Regression} + SQ_{Residuals}$

This partitioning of the total sum of squares provides us with the regression sum of squares and the error or residual sum of squares. $\mathbf{SQ_{Residuals}}$ will be minimized by the least-square method (Equation 3.1). In case of a perfect regression no residuals occur and $\mathbf{SQ_{Residuals}}$ is equal to zero. Thus, the quality of description by the model is increasing with an increasing amount of $SQ_{Regression}$ on the total value SQ_{Total}. The **coefficient of determination R^2** (commonly also **goodness of fit**) is defined by:

Equation 3.5: $$R^2 = \frac{SQ_{Regression}}{SQ_{Total}} = 1 - \frac{SQ_{Residuals}}{SQ_{Total}}$$

R^2 ranges between 0 and 1. 1 denotes a perfect correlation, e.g. all points lie on a straight line. When there is no correlation R^2 is zero. 0.75 indicate a rough, but stable relationship with still practical importance.

An important value is also the root square of the coefficient of determination and is called **correlation coefficient R**. R ranges between 0 and 1. 1 denotes once more a perfect correlation.

Correlation coefficients are simple to calculate. Thus, they are all too easily misinterpreted. The data points and the regression line as well as the residuals must always be plotted (see Figure 3.2). Otherwise straight lines might be deduced although the curve is sufficiently gentle.

If a model is expanded (e.g. by taking into account interactions), the quality of the model and thereby the correlation coefficient can be improved. Finally a value of R = 1 can be obtained by simply fitting a regression equation that includes via a polynomial model ($y = a_0 + a_1 \cdot x + a_2 \cdot x^2 + a_3 \cdot x^3 + a_4 \cdot x^4 + \ldots$) as many terms as there are observations (**over-fitting**). Finally the correlation coefficient will be large, even there is no true relationship between the variables and a linear relationship would describe the correlation much better. It doesn't make sense

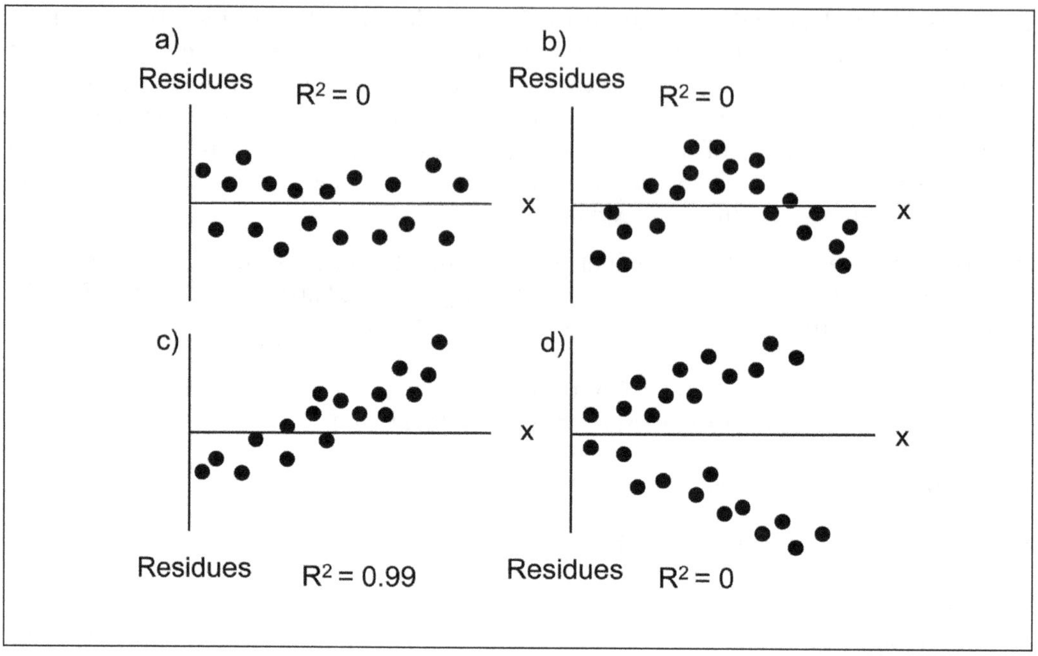

Figure 3.2: Residues as function of the factor: a.) ideal behaviour, b.) non-linear behaviour, c.) linear residues – wrong model, d.) inhomogenity of the residues

to deal with a model, which describes every fluctuation. In addition polynomial models have drastic deviations at the edges of an experimental area (see Figure 1.31 c) [1]. As already *Einstein* said: Everything should be made as simple as possible, but not simpler.

Thus, very often a second parameter **Q^2, goodness of prediction** (often also **cross-validated R^2**) is calculated to avoid the phenomenon of over-fit.

Equation 3.6: $$Q^2 = 1 - \frac{SQ_{PRESS}}{SQ_{Total}}$$

with

Equation 3.7: $$SQ_{PRESS} = \sum_{i=1}^{N} (y_i - \hat{\hat{y}}_i)^2$$

Q^2 ranges between $-\infty$ and 1. Values above 0.5 describes a good and above 0.9 an excellent model.

PRESS stands for **pr**ediction **e**rror **s**um of **s**quares and is an indicator how correctly the model for the experiment is able to predict the responses. In principle the calculation is the same as in case of R^2. In case of Q^2 we select an observation – for example i. Now, we fit the regression model (**reduced model**) to the remaining n - 1 observations and use this equation to predict the withheld observation y_i. Denoting this predicted value $\hat{\hat{y}}_i$, we may find the prediction error of point i as $y_i - \hat{\hat{y}}_i$. The prediction error is often called as PRESS residual (see Chapter 3.3.4). The procedure is repeated for each observation, producing a set of n **PRESS** residuals. Now the sum of squares of theses residuals is formed. SQ_{PRESS} is always larger than $SQ_{Residuals}$. Thus, R^2 is always larger as Q^2 and must exceed Q^2 by more than 0.2 to 0.3. For bad predictive models, Q^2 can assume even negative values when SQ_{PRESS} is greater than SQ_{Total}. R^2 and Q^2 must be evaluated always together and an optimization has to be performed very balanced. In case of large R^2 but small Q^2 low influence terms of the model should be eliminated.

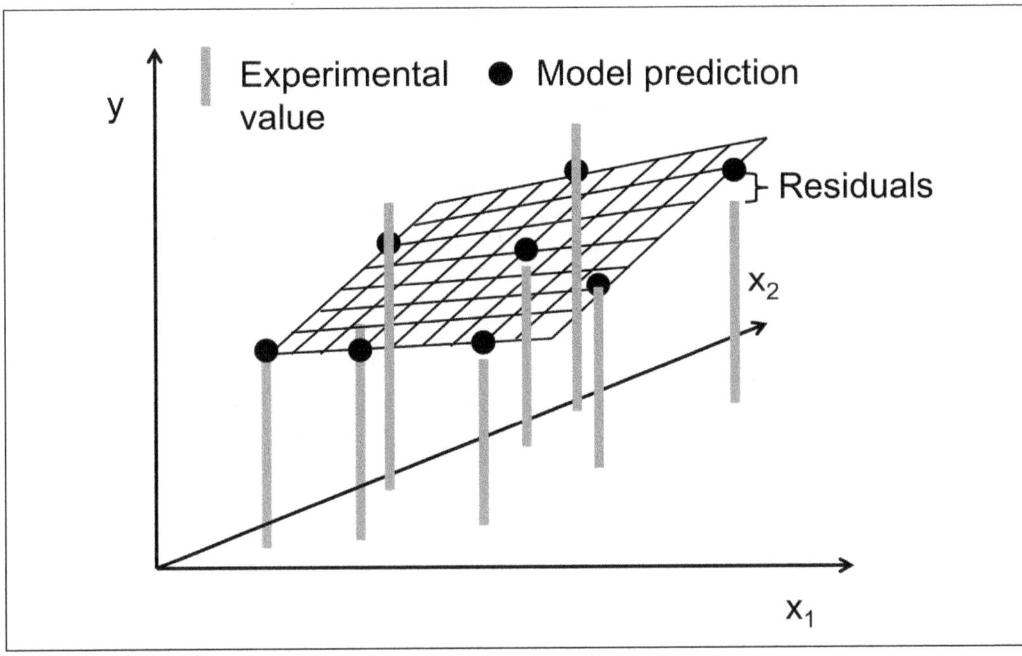

Figure 3.3: Principle of the regression analysis in case of response surface method[26]

The method of regression analysis can also be applied to the general equation relating one target value with several independent variables (= **multiple linear regression**) [2]. It is also possible to analyze non-linear systems, but in this case it is much easier to use software packages (see Chapter 5). Without computer software it would be incredibly difficult to fit all the coefficients in the models. In case of more than one variable so called response surfaces can be derived (see Figure 3.3). This is the basis for the **response surface methodology** (RSM), which is useful for the analysis and optimization of problems with several variables (see Chapter 4). In addition RSM has become also more prevalent in verifying the reliability of a regression model prior to only the conversion to response surface plots.

▶ ***Basic principle:*** To evaluate the goodness of fit ("explained variation") the coefficient of determination R^2 is used, which ranges between 0 and 1. 1 denotes perfect correlations, 0.75 indicates a rough, but stable and useful relationship and 0 corresponds to no correlation at all. It measures how well we can reproduce current runs. An important value is also the root square of the coefficient of determination and is called correlation coefficient R. To overcome the phenomenon of over-fitting a second parameter Q^2, goodness of prediction ("predicted variation") is calculated. Q^2 ranges between $-\infty$ and 1. Values above 0.5 describes a good and above 0.9 a perfect model. It uncovers how well we can predict new experiments. R^2 is always larger as Q^2 and must exceed Q^2 by more than 0.2 to 0.3.

3.3.2 *Confidence intervals for the model parameters*

Deviations, which the model cannot explain should be in average zero. It is desirable that each point is as close as possible to the regression line. This variance of the data points around the regression curve (= residuals), no matter if generated by a single measurement or replicate measurements (together with the calculation of the average value), affects a certain variability in the model. The standard deviation of the residuals is a kind of indicator for this issue and the value is transferred to the model. In addition, each data point,

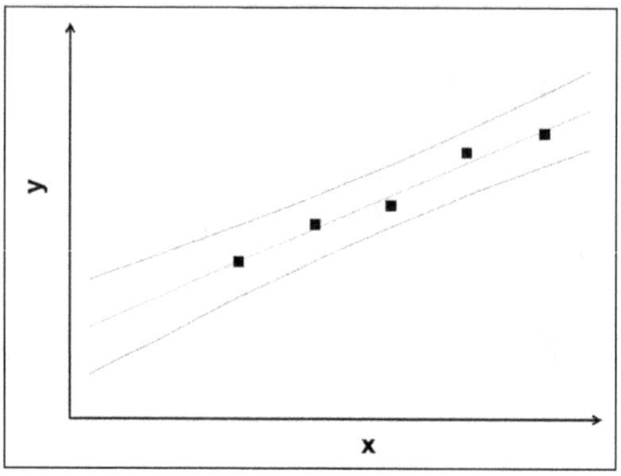

Figure 3.4: General form of the confidence limits (95 %) for a linear regression

which is usually an average of several replicates, has a certain confidence interval. Also this noise is transmitted to the slope and intercept. Although the calculation is not explained in detail, more information can be found in Appendix 7. The basics for this procedure can be found in Appendix 4 including the correction with the so called t-factor. Finally a confidence interval for the regression curve can be calculated (see Figure 3.4).

The confidence intervals for the model parameters enable also to eliminate not significant terms from the model. Of course this is only possible as long as no technical reasons said against it. Thus a **reduced model** is generated which still describes the reality.

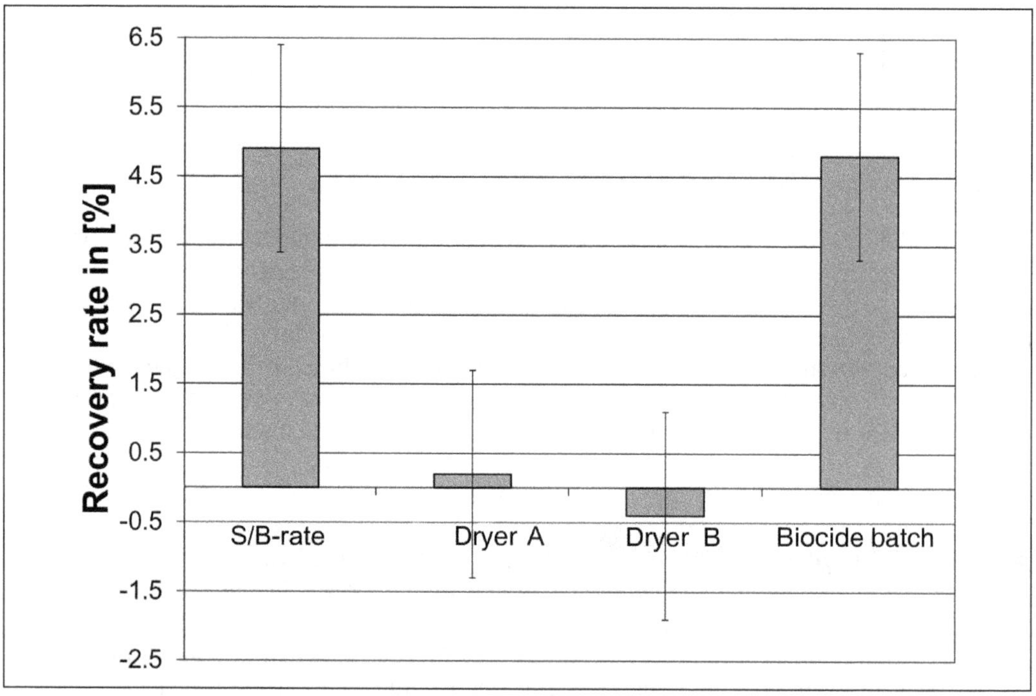

Figure 3.5: Plot of the model coefficients (= effect/2) with error bars, see example in Chapter 2.3.2

The confidence interval of slope and intercept are also equal to the confidence intervals of the effects. The model coefficients are still equal to effect/2 (see Figure 1.28). Thus, some software packages plot the model coefficients instead of the effects in the effect-diagram. Be aware of that, because very easy errors occur due to this fact, because these diagrams doesn't vary from a qualitative point from the *Pareto* diagram (see e.g. Figure 3.5). In addition, confidence intervals always include the random error of the experimental data, but still also the lack of fit due to wrong regression models. Thus, significance is also a function of the goodness of fit, as shown in Chapter 2.3.2 and 3.5.1.

3.3.3 *Basic principles and standard assumptions for regression analysis*

For a successful application of the regression analysis non of the basic principles and standard assumptions should be violated:

- The selection of model influencing factors must be very careful. **Missing factors** causes poor models. Reality should be described perfectly, but be aware of **over-fitting**.
- Reality is described by a simple model. **Non linearities** and interactions have to be taken into account. However, **reduced models** with quadratic or even cubic terms can only express the real characteristics if the experimental design has the right configuration. **Models** are intimately **linked** with the **experimental design**. Thus, it has to be planned already in a previous step in which experimental area the model should fit.
- **Enough data** must be present. Usually twice as much data points as model parameters are adequate. Even more data points end up in better predictions.
- The data set has to be evaluated regarding to the **statistical requirements** (e.g. probability plot, histogram, residuals N-plot, see Chapter 3.3 and Appendix 3). Primary raw data has to be evaluated for reproducibility (variance). Sometimes the model quality benefits from **data transformation**. Small variability is an advantage (display the measured responses for all experimental runs in a replicate plot, see Figure 2.28). It is recommended to make a data preparation before calculating the model. **Time trends** and **outlier** analysis increase the chance for good models (see Chapter 3.3).
- Every model has to be analyzed on its **plausibility**. In addition, be careful with very simple models based on two or three data points (see Figure 1.31). There is a high chance to run the risk of overseeing non-linear behaviour. Thus, curvature causes poor models.
- Check the model assumptions and **evaluate the model** equation as well as the **regression coefficient**, the **goodness of prediction** and the **residuals** (see Chapter 3.3). R^2 should be > 0.75, indicating a relationship of practical importance and Q^2 should be at least above 0.5, describing a correct model.
- Models have always only a **local explanatory power**. Inside the analyzed area interpretations are possible. An extrapolation, meaning a prediction of values outside the analyzed area is forbidden (see Figure 3.6).

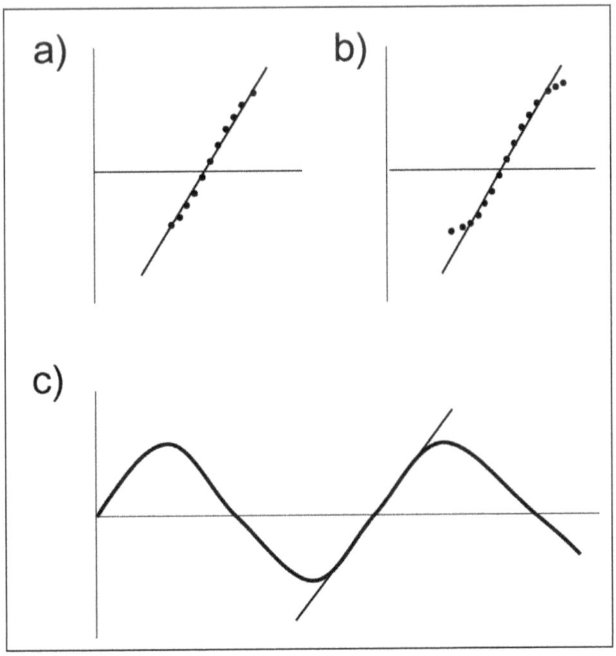

Figure 3.6: Extrapolation is forbidden: a) A good model in the considered area; b) Expanding the area: it could be better!; c) Whole area: Model ≠ Reality!

Data analysis by regression is essential in case of designs of higher levels, because they cannot be analyzed only by the confidence interval. However, in case of qualitative factors and multilevel designs also models have their limits. In this case analysis of variance (ANOVA) is the right method (see Chapter 3.5).

3.3.4 Example determination of the hardener with DSC

The glass transition temperature T_g of a solventborne 2K-polyurethan clear coating is determined by differential scanning calorimetry (DSC) as a function of the amount of hardener (each data point with 3 replicates, Table 3.1). With this data a calibration curve for the analysis of the amount of hardener in the coating is calculated.

Data analysis (see Appendix 7) ends up in the following data set and calibration curve (Figure 3.7):

Working area ranges from 10 % to 30 %
Number of calibration samples: $N_K = 5$; $N_a = 1$
Grand average calibration concentration: $x\text{-}\bar{x} = 20$ %
Grand average of the response: $y\text{-}\bar{y} = 60.6$ °C
Intercept: $a_0 = 19.7$ °C
Slope: $a_1 = 2.0$ °C/%
Standard deviation of the slope: 0.15 °C/%
Standard deviation of the intercept: 3.46 °C
$Q_{xx} = 250$
$S_{x0} = 1.15$ %

Confidence interval at the borders of the experimental area:

VB (x = 10 %) = 4 %
VB (x = 30 %) = 4 %

Regarding the regression coefficient also the square sums are calculated (see Table 3.2):

$$Q^2 = 1 - \frac{SQ_{PRESS}}{SQ_{Total}} = 0.97$$

$$R^2 = \frac{SQ_{Regression}}{SQ_{Total}} = 1 - \frac{SQ_{Residuals}}{SQ_{Total}} = 0.98$$

Table 3.1: Linear regression - experimental data for the determination of the glass transition temperature as a function of the hardener

Data point	Amount of hardener in [%]	T_{g1} in [°C]	T_{g2} in [°C]	T_{g3} in [°C]
1	10	40	38	42
2	15	52	55	49
3	20	58	58	56
4	25	73	70	75
5	30	81	80	81

Table 3.2: Calculation of R^2 and Q^2 with $\bar{y}$ = 60.6 °C for the glass transition temperature

x_i	y_i	$\hat{y}_i$	$\hat{\hat{y}}_i$	$y_i-\hat{\hat{y}}_i$	$(y_i-\hat{\hat{y}}_i)^2$	$y_i-\bar{y}$	$(y_i-\bar{y})^2$	$y_i-\hat{y}_i$	$(y_i-\hat{y}_i)^2$
10.0	40.0	39.7	40.1	-0.1	0.0	-20.5	421.6	0.3	0.1
15.0	52.0	49.7	49.8	2.2	4.8	-8.5	72.8	2.3	5.3
20.0	57.3	59.7	61.3	-4.0	15.7	-3.2	10.2	-2.4	5.6
25.0	72.7	69.7	69.8	2.9	8.2	12.1	147.2	3.0	8.8
30.0	80.7	79.7	82.3	-1.7	2.8	20.1	405.4	1.0	0.9
Sum					31.6		1057.2		20.7
					= SQ_{Press}		= SQ_{Total}		= $SQ_{Residuals}$

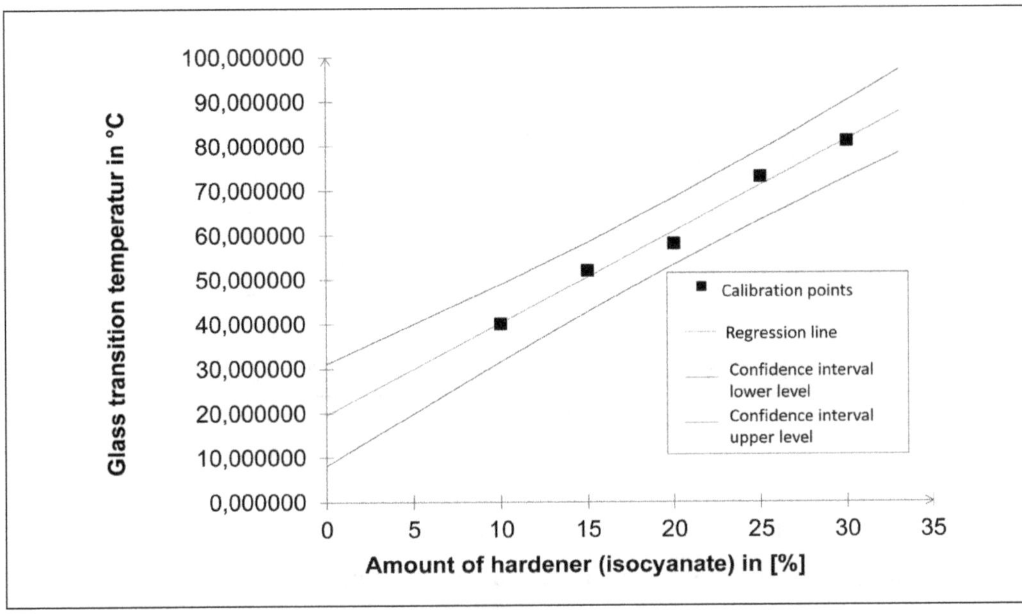

Figure 3.7: Linear regression for the determination of the amount of hardener via glass transition temperature

The glass transition temperature of the clear coating is increasing with the amount of hardener (see Figure 3.7). The correlation has a clear linear character and replicates will be at the boarders of the calibrated area (10 % and 30 % hardener) with 95 % probability inside a confidence interval ± 4 %.

3.4 Residual diagnostic – what does the deviations mean?

The analysis of residuals is essential to evaluate the quality of the model, because a lot of the consecutive aspects are also basic requirements for several evaluation steps (e.g. normal distribution). Please be always aware of the following issues:

- Is there a consistence between data and model? Is the model describing the data in a good manner („**observed versus predicted plot**")? Residuals should be small in general. A plot of the predicted values vs. the experimental data points is very helpful in this step (see Figure 3.9). The tighter the scatter of points around the straight line, the stronger the model. Outliers are identified very easy and the picture gives a good overview about the already present model quality. Be aware of non-linear behaviour, which is not considered in the model. This might be indicated by a curved structure.
- Are there any trends or time-dependent effect? If we plot the residuals vs. time or against the order of experiment, no structure should result. In other words: knowledge about any data point must not enable the prediction of the following experiments (see Figure 3.8 and 3.11).
- Is there any correlation between the standard deviation (variance) of the data points and the response itself? Check, whether the residuals are proportional to the predicted model values (see Figure 3.2 and 3.10). If the correct model has been used, the residuals remain roughly uniform in size as y_i increases.
- Testing for normality of distribution by plotting the data on a normal probability paper (**Residual N-plot**). Please find details in literature [3, 4] or Appendix 3. It is also very easy to see whether there are outliers in the dataset.

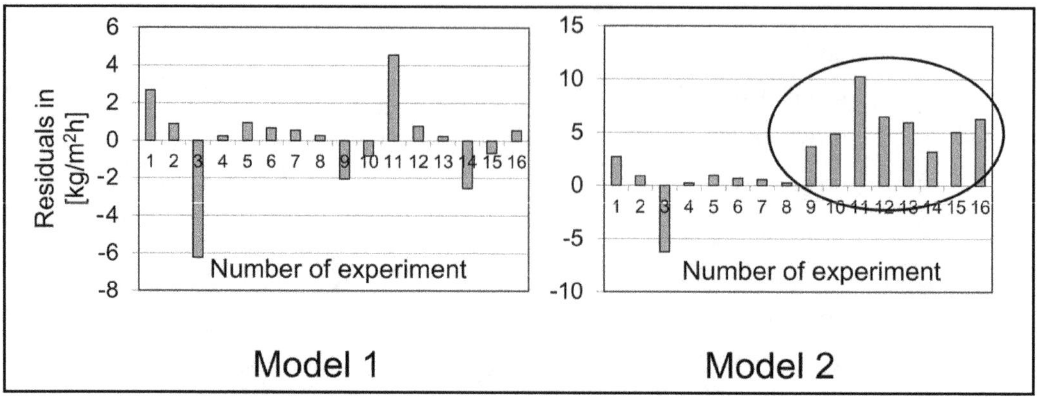

Figure 3.8: Residual diagnostics for both models

- Testing small variability in the replicates (replicate plot, see Figure 2.28 ; other possibilities are histogram or *Box*-plots, see Appendix 3)

Example: Modeling – drying rate of a waterborne coating

The drying rate of a waterborne coating (kg/m^2h) are analyzed regarding the influencing factors temperature A, humidity B, air velocity C, nozzle drying or circulating air (= D). Response and target value is the time to get a non-adhesive surface. Two different models are fitted to the data set and compared by residual diagnostics:

Model 1: $W(A,B,C,D) = a_0+a_1 \cdot A+a_2 \cdot B+a_3 \cdot C+a_4 \cdot D$ (linear)

Model 2: $W(A,B,C,D) = a_0+a_1 \cdot A^2+a_2 \cdot B+a_3 \cdot C+a_4 \cdot D$ (quadratic part)

No structure is obvious in the residuals of model 1 (see Figure 3.8). Residuals are small and no trend is present. Time-dependent effects are very likely in case of drying experiments because humidity could be accumulated in the setup. However, in this experiment this seems not to be the case. To overcome time-dependent effects blocking could be helpful.

In contrast to that, model 2 has after experiment 9 always a positive deviation of the model to the experimental data. Thus, no correct description of reality can be achieved by the model 2.

Example: Chemical resistance against creams and fat in case of a 2K-top coat

In this example a waterborne 2K-topcoat should be optimized and the product is mainly used inside of vehicles. Thus, in addition to optical aspects also surface haptics and the resistance against creams (e.g. sun cream, hand cream) are essential parts of the customer requirements. Various coating components (e.g. the additive package) and the curing degree are analyzed in an experimental screening design regarding their influence on the chemical resistance against creams. A visual evaluation of the surface takes place after exposure to a certain sun cream. Afterwards also an adhesion test is done by cross-cut test. Adhesion is classified by school grades (from 0 to 5 in integers) and the effects on coating formulations are described in a model.

It is obvious from the residual diagnostics that this semiquantitative and subjective acquisition of the response values via school grades doesn´t ends up in an optimal mathematical model (see Figure 3.9, 3.10, 3.11 and 3.12). Significant deviations from the basic principles of modelling are present. However, in general qualitative evaluation does not end up in weak correlations, models and statistics. If any of the basic principles like objective evaluation, not too fine scale-steps, regular scale of grades, etc. are not violated (see Chapter 2.6), statistical analysis produces highly significant models also in case of qualitative data. In this exam-

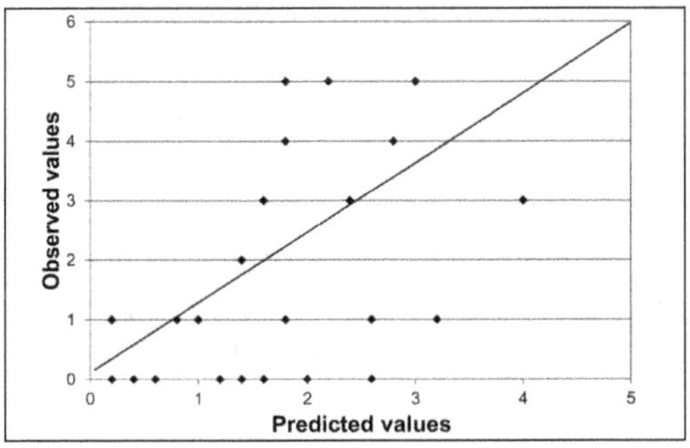

Figure 3.9: Residual diagnostics - observed vs. predicted values, in case of the surface evaluation after exposure to sun cream (school grades from 0 to 5)

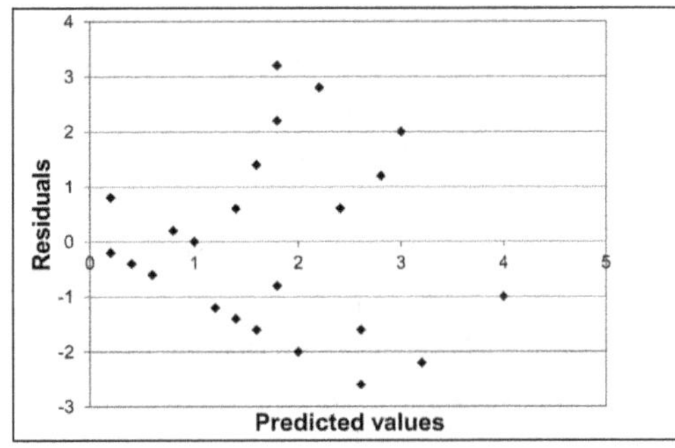

Figure 3.10: Residual diagnostics - residuals vs. the predicted values, in case of the surface evaluation after exposure to sun cream (school grades from 0 to 5)

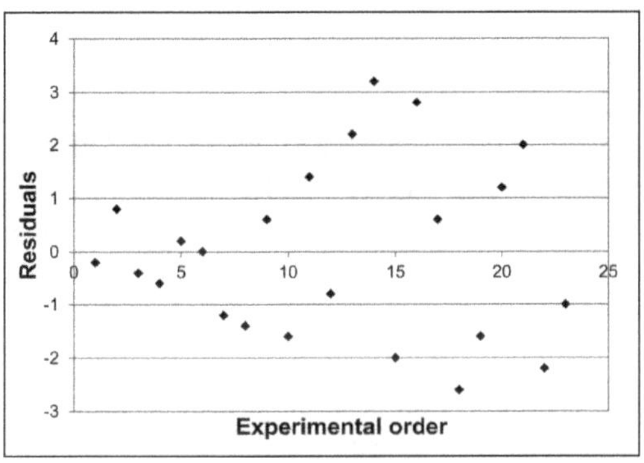

Figure 3.11: Residual diagnostics - residuals vs. the experimental order, in case of the surface evaluation after exposure to sun cream (school grades from 0 to 5)

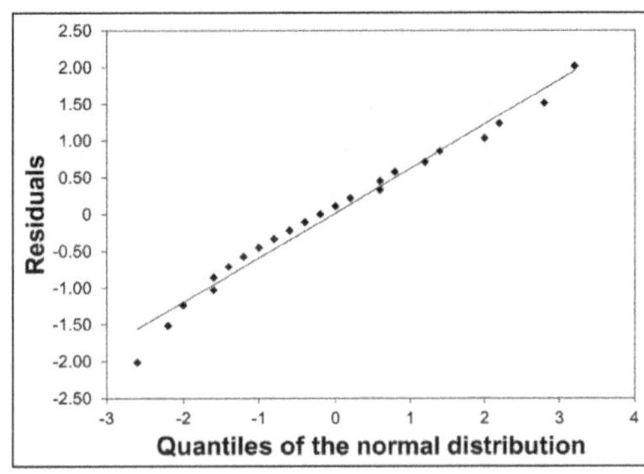

Figure 3.12: Residual diagnostics - residuals vs. the quartiles of the normal distribution (residual N-plot), in case of the surface evaluation after exposure to sun cream (school grades from 0 to 5)

ple, the variance is in addition increasing by a worse reproducibility of the results. To avoid this, much more experiments and replicates are necessary. In case of the above mentioned example already a worse data set is present in principle and a serious analysis cannot be performed also with more replicates. However, even if the goodness of prediction is worse in such cases, DoE is able to give at least an information about effects or a prediction of a trend in a clear and reproducible manner.

Similar problems due to worse reproducibility and not objective evaluation take place also with the classical approach. However, very often the data set of the common procedure is much larger and very confusing.

3.5 Analysis of variance - how certain we can feel?

3.5.1 Introduction

Analysis of variance (frequently abbreviated to **ANOVA**) is an extremely powerful tool to separate and estimate the different causes of variation [5, 6]. Thus, it is possible to separate any variation on the response value which is caused by altering one factor, from variations due to random error (significant difference between two values like in Chapter 1.6). This way of data analysis is very common in software packages and is used also for simple designs and for the comparison of two means. In these cases the same result is obtained by ANOVA and the comparison of two means (which was preferred in the book until this chapter). However, in practice there are often more than two means to be compared. In addition, in case of multi-level designs regression analysis is limited to quantitative data. ANOVA can be applied also to qualitative data.

In order to estimate how well the experimental points fit to the linear regression model given by Equation 1.10 in Chapter 3.3 (regression analysis), the quantity **SQ_{Total} (= total sum of squares, TSS)**, the sum of squares of the deviations from the mean $\bar{y}$ for the individual y_i values (Equation 3.2 and 3.8), was defined.

Equation 3.8: $$SQ_{Total} = \Sigma_{i=1}^{N} (y_i - \bar{y})^2 - \frac{1}{N}$$

with $\bar{y} = \frac{1}{N} \Sigma_{i=1}^{N} y_i$, the mean of all response values and N the total number of data points

This parameter is also very helpful in analyzing general experimental designs, because, in analogy to regression analysis, variance is based on random errors and on real effects which can be explained. Thus, SQ_{Total} will be once more partitioned (see Equation 3.9), but now into a sum of squares of differences between the level average value (average of all observations at a certain level) and the **grand average over all observations SQ_{Effect}**, plus a sum of squares of the differences of all observations within a certain level from the level average **SQ_{Rest}** (average of all observations at a certain level). These values SQ_{Total}, SQ_{Effect}, SQ_{Rest} will be calculated for each variable itself.

Equation 3.9: $$SQ_{Total} = SQ_{Effect} + SQ_{Rest} = N_G \cdot \Sigma_{j=1}^{M} (\bar{y}_{Group,j} - \bar{y})^2 + \Sigma_{j=1}^{M} \Sigma_{i=1}^{N_G} (y_{ij} - \bar{y}_{Group,j})^2$$

with M = the number of groups (each group has the same number of data points)
N_G = the number of data points in one group
y_{ij} = the data point i in group j
$\bar{y}$ = the grand average of all data and
$\bar{y}_{Group,j} = 1/N_G \Sigma_{i=1}^{N_G} y_{ij}$ the average value of all data in group j

To get an idea about this splitting, think about the determination of the sum of distances from all data points to the grand average over the complete data set. This can also be done in two steps. First of all by formation of the sum of squares of the difference inside the groups referring to the particular average value of the group $\bar{y}_{Group,j}$. Afterwards the sum of squares of the difference of each average value of the group $\bar{y}_{Group,j}$ to the grand average $\bar{y}$. To end up with the total sum of squares, the sum of squares of each average value of the group $\bar{y}_{Group,j}$ to the grand average $\bar{y}$ has to be multiplied with the total number of data points in one group N_G (see Equation 3.9 and 3.12). Otherwise the distance of each data point to the grand average, which is essential in the calculation of SQ_{Total}, is not taken into account. The following example should describe the calculation more in detail.

Example: Affecting the structural viscosity of a waterborne pigmented coating

A coating formulator is conducting an experimental to study the structural viscosity of a waterborne pigmented coating. The factors he initially regards as important are the thickener, a surfactant and the solvent. Target value is the ratio of the elastic modulus (G`) to the viscous modulus (G``) from an oscillatory-rheological measurement (see Table 3.3).

Great average ratio G`/G`` over all data: 1.21

Thus the total sum of square is calculated by:

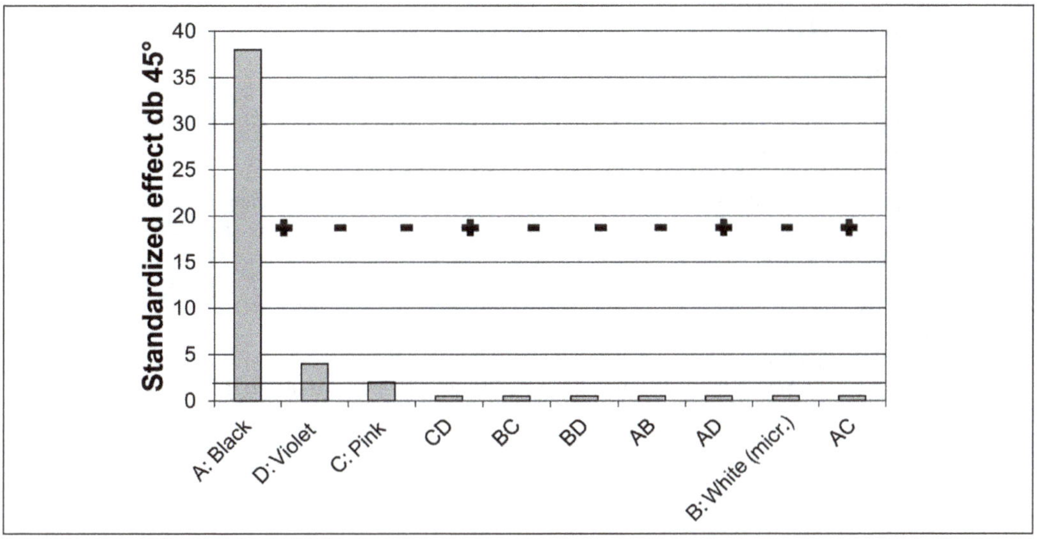

Figure 3.15: Pareto-Diagram regarding the example colour measurement of a blue-metallic base coat

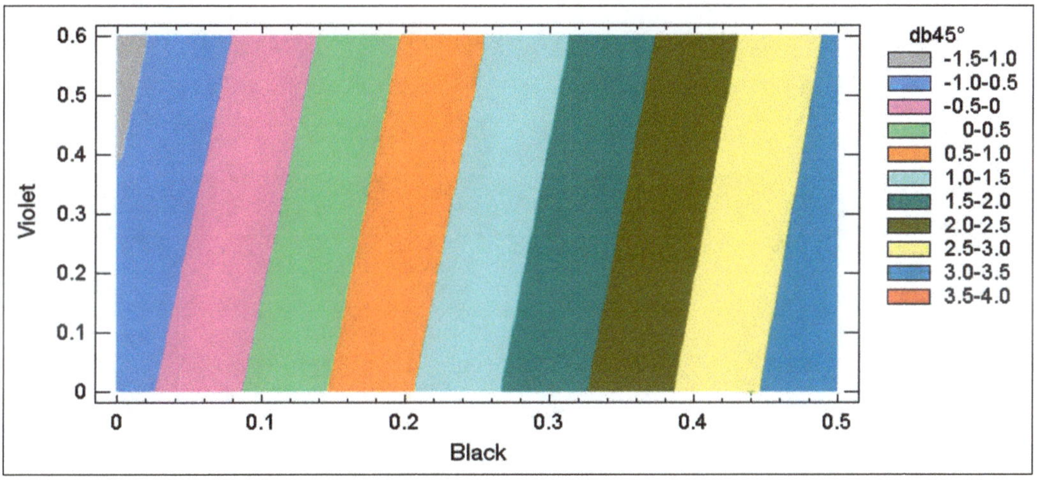

Figure 3.16: Response-surface as contour plot (model 2) regarding the example colour measurement of a blue-metallic base coat; white (micr.): 0.44 %, pink: 0.68 %

Table 3.11: Analysis of variance of db 45°-model 2 regarding the example colour measurement of a blue-metallic base coat

Factors	Sum of square	f	Calculation including the degree of freedom	F-ratio	p-value
A:Black	57.881	1	57.881	4108.27	0
B:White	0.0034	1	0.0034	0.24	0.6574
C:Pink	0.191	1	0.191	13.58	0.0346
D:Violet	0.547	1	0.547	38.84	0.0083
Lack-of-fit	0.355	12	0.029	2.10	0.2952
Random error	0.0423	3	0.014		
Total (Corr.)	58.503	19			

$R^2 = 99.3207$ %

3.6 *References*

General literature references can be found in the Appendix 9.

[1] K. Backhaus, B. Erichson, W. Plinke, R. Weiber, Multivariate Analysemethoden. Eine anwendungsorientierte Einführung, Berlin, 11th Edition 2006
[2] R. Henrion, G. Henrion, Multivariante Datenanalyse, Springer Verlag, Berlin, 1995
[3] S. Chatterjee, B. Price, Praxis der Regressionsanalyse, Oldenbourg Verlag, München, 1995
[4] K. Siebertz, D. van Bebber, T. Hochkirchen, Statistische Versuchsplanung, Springer Heidelberg 2010
[5] J. Hartung, B. Elpelt, Multivariate Statistik, Oldenburg Verlag, München, 1992
[6] E. Kreyszig, Statistische Methoden und ihre Anwendungen, Vandenhoeck & Ruprecht, Göttingen, 4th Edition, 1973
[7] W. Funk, V. Dammann, G. Donnevert, Qualitätssicherung in der Analytischen Chemie, 2nd Edition, Wiley-VCH, Weinheim, 2005
[8] R. Schlittgen, Einführung in die Statistik, 10th Edition, Oldenbourg, München, 2003
[9] W. Kleppmann, Taschenbuch Versuchsplanung - Produkte und Prozesse optimieren, 6th Edition, Hanser, 2009

4 Parametric optimization and sensitivity analysis – finding a needle in the haystack

The objective in **optimization** is extracting in-depth information about the few dominating factors selected in the screening phase. In addition, until now many factors were investigated in a few runs, but in optimization the reverse is true, because few factors are explored in many experiments [30]. Beside the already in previous chapters emphasized cause-and-effect correlations this stage focus on the location of new promising experimental points. In addition, this stage is used to predict where to do the next experiment, which hopefully will correspond to, or be in the vicinity of, the optimal point (e.g. maximum or minimum response) [30]. Due to simulation and analysis of the influencing factors and noise factors also more **robust products and processes** can be achieved (sensitivity analysis). Just optimizing a system does not consider this aspect and the optimum can be very sensitive against variations. Before the release of an almost finished product, such tests are usually carried out as a last check to ensure quality. Depending on the question several methods are available. Finding the needle in the haystack might start now.

4.1 Strategies for optimization – how we can do it better

The aim of all different optimization strategies is to reach the optima with a minimal amount of steps from the starting point [1-3]. Three methods are most important and are described in the following chapters:

- Method of the steepest ascent/descent (so-called **gradient technique**)
- EVOP-method
- Simplex-method

All three methods are part of the **response surface method (RSM)**. Response surface methodology is a collection of mathematical and statistical techniques for empirical model building. By careful design of experiments, the objective is to optimize a response (output variable) which is influenced by several independent variables (input variables). Fitting a model to the experimental data is fundamental in RSM (see Chapter 3.3). RSM has a sequential nature for optimizing a process by moving to the optima due to modifying the experimental design. Such techniques are very useful also to re-adjust factor setting and move the experimental design. The ideal case is of course if one of the performed screening experiments fulfills the experimental goals. Sometime, however, it will turn out difficult or even impossible to locate an interesting area inside the experimental domain. Thus, finding the optimal region has to be done as bridge between screening and real optimization. Especially the simplex technique is predominantly used for extrapolation purposes, that is, for prediction outside the investigated experimental region. This procedure may be used for interpolation as well.

4.1.1 *Method of the steepest ascent/descent*

The process of optimization can be visualized in terms of a blindfolded person climbing a hill with the task of finding the summit. Under these circumstances a possible approach is to walk always in the direction in which the hill is steepest. A stream is also always searching for the fastest way downhill and is following the **steepest ascent**.

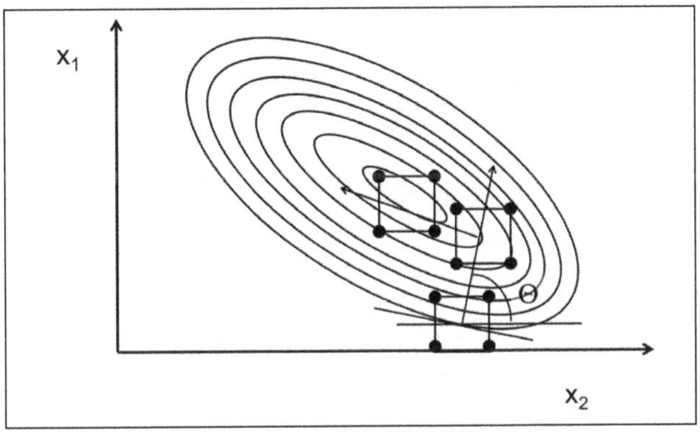

Figure 4.1: Method of the steepest ascent; the arrow indicates the direction of the steepest ascent in the experimental design

The same procedure can be applied in case of experimental design. Considering a factorial experiment at two levels in a first step with already calculated main effects and the model function. The main effects of both factors x_1 and x_2 give already a hint in which direction the maxima will be located. In case of a positive main effect of both variables, the maxima will be located to the right and above the actual experimental area. If the effect of x_1 is greater than that of x_2, we must search the maxima more above the starting point then in a region to the right. More precise, it is possible to derive an angle Θ for the steepest ascent from the ratio of the main effect or the slope b of a model (Equation 4.1 and Figure 4.1).

Equation 4.1: $$\tan\theta = \frac{HE(x_1)}{HE(x_2)} = \frac{b_2}{b_1}$$

4.1.2 *Box-Wilson's method*

Bottom-up to the method of steepest ascent a procedure can be generated to determine the optimal parameters by combination with the DoE approach. The ***Box-Wilson* method** is starting inside a certain experimental area with an experimental design and the method of steepest ascent. The next step is the optimization by further experiments in the direction indicated by the steepest ascent. This procedure continues until no further progress is possible due to worse response values. Afterwards another factorial experiment is carried out to determine the new direction of steepest ascent [28] (Figure 4.1).

Local and close positioned optima will be found very easy. However, it is evident that finding the optima depends strongly from the starting point and hardly any global maxima or minima are found. In this case it is indispensable to start from different points.

4.1.3 *EVOP-Method (evolutionary operations)*

This method is a special version of the method of steepest ascent, which is very often used in manufacturing processes [4]. **EVOP (evolutionary operations)** is intended to introduce small changes in the process variables during normal production flow. The levels are so small, that no difference is obvious to the customer and no interruption of the production process is necessary. Thus, effects are small, but due to the huge number of samples the statistical approach is still possible.

4.1.4 Simplex-method

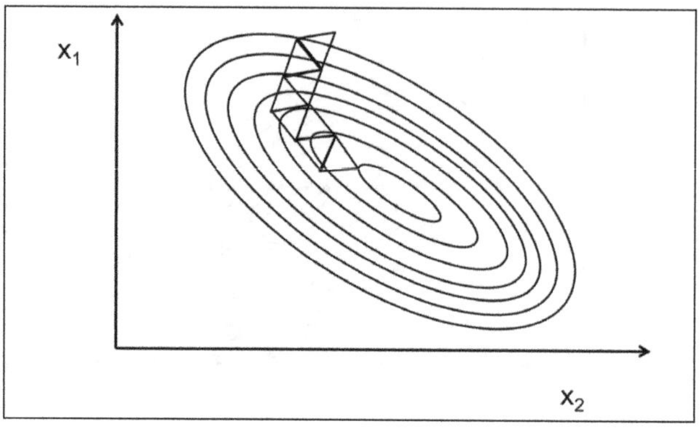

Figure 4.2: Simplex-method

In case of two factors this optimization procedure is performed by a so called simplex design, which will be a triangle. In the first experiment the response is measured at every combination of factor levels given by the vertices of the **simplex**. The worst response in this case is the trigger to continue at the point which i•s the reflection of the triangle. Mirror the worst experiment through the line connecting the two best experiments [30]. The reflection process is repeated until no further progress is possible and a worse response is resulting. Be aware, that this method has nothing to do with the simplex designs in Chapter 2.5 and this is only a simplified picture of how the method works [24]. The method tends to find only local minima. The global optimum, rather than only a local one, is provided by starting in differnt regions and checking, whether the same optimum is optained.

4.1.5 Further optimization methods

Today, several other methods are used to improve the finding of the global optima. Very common are evolutionary algorithm [5, 6, 21]. The most popular type is the **genetic algorithm**. The evolution usually starts from a population of randomly generated individuals, which determine the response values at their local position. All these data are collected in a so called fitness function, which will be optimized during the procedure. Depending on the value of the fitness function the next steps are defined and this procedure is repeated until a certain stop criterion has been reached. This method is very effective in case of large experimental areas, but needs a lot of computer power. However, very often in reality the search criteria are not well defined and the aim is to find regions with as many local minima and maxima as possible. In this case **cluster analysis** [7, 27], **neuronal nets**, sometimes called also **artificial neural nets** or ANNs [8-11, 20, 23, 26], or **fuzzy logic** [11-13] are used.

4.2 Multiple responses

Typically a coating formulation has to fulfil several requirements. Thus, simultaneous consideration of multiple responses is reality (see also Chapter 1.4.2.4) [23]. Only one response has to be optimized rarely! Very often it is difficult to find a set of operating conditions that optimizes all responses (**sweet spot**). However, they are located at completely different positions. Thus, at least we should try to keep them in desired ranges by finding a compromise. In principle two possible approaches exist [15]:

- First of all an appropriate model for each response has to be built. Starting with the dominant variables always two responses are optimized subsequently. Contour plots are very helpful using the traditional overlay technique. The next step takes into account that some responses are already optimized and the relevant factors should be hold constant and other responses are not so important. This approach is relatively straightforward when there are only a few variables. When there are more than three design variables, overlaying is impossible and a lot of trial and error is required.

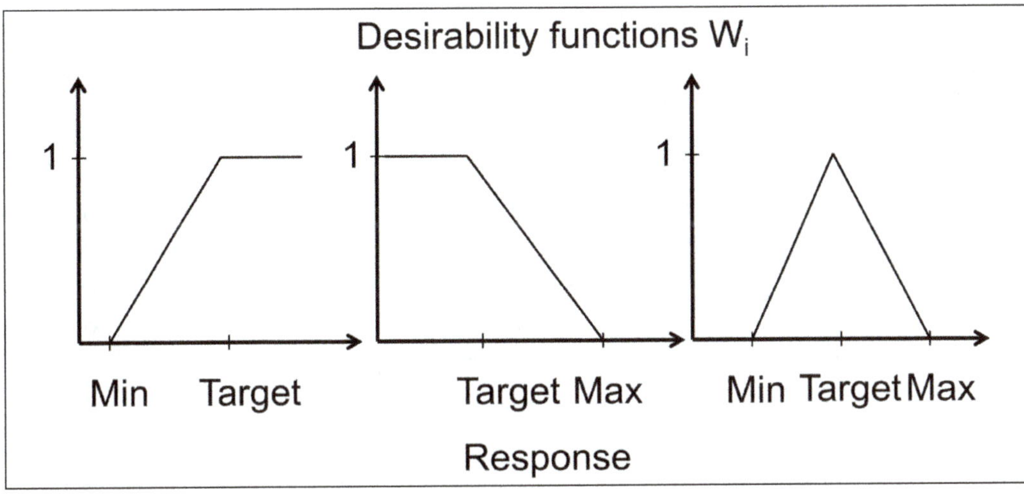

Figure 4.3: Several desirability functions

- Another useful approach is the **desirability functions**. First of all each response is converted into an individual desirability function W_j, that varies from 0 to 1. The function equals to 1 if the response is at its target. Afterwards the operating conditions are chosen to maximize the function given by the geometric average (Equation 4.2):

Table 4.1: Factor settings regarding the example multiple response optimization of blocking and film formation

Factors	Low limit in [%]	High limit in [%]
Water	0.0	6
Hydrophilic Coalescent	0.0	6
Hydrophobic Coalescent	0.0	6

Equation 4.2: $W_{total} = (W_1 \cdot W_2 \cdot \ldots \cdot W_k)^{1/k}$

W_{total} is 0, if already one $W_i = 0$. W_{total} is only equal to 1, if all W_i are equal to 1 (see Figure 4.3).

Example: Multiple optimization of blocking and film formation in a clear coat

Two responses – blocking and film formation – of a clear coat will be optimized depending on the solvent composition. A hydrophilic and hydrophobic coalescent as well as water are mixed in several ratios, but the total amount is always 6 % in the coating formulation (mixture design). The response is on base of qualitative school grades: film formation (from 0 = perfect film formation to 3 = cracks) and blocking (from 0 = no blocking to 3 = strong adhesion). We search now for a solution that maximizes overall desirability: film formation < 1.0 and blocking < 1.0. Blocking has

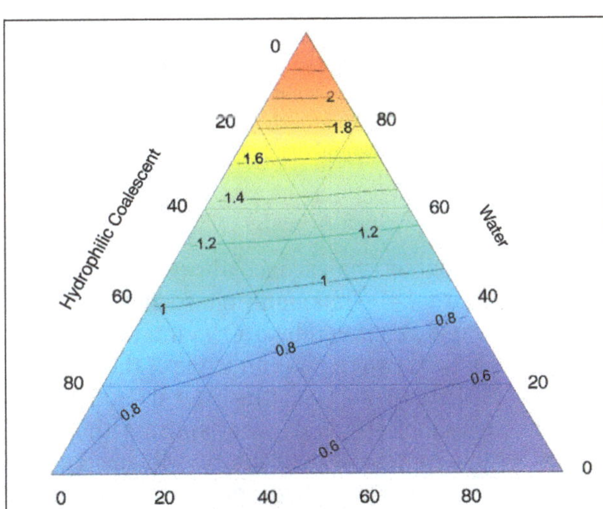

Figure 4.4: Response surface for the film formation

Table 4.2: Design matrix regarding the multiple optimization of blocking and film formation in a clear coat

Name of the experiment	Number of experiment	Amount of hydrophobic coalescent	Amount of hydrophilic coalescent	Amount of water	Blocking	Film formation
1	5	0.33	0.33	0.33	0.7	0.3
2	9	0.33	0.67	0	1	0
3	1	0.67	0.33	0	0.7	0
4	11	0	0	1	0	2
5	2	1	0	0	0	0.5
6	12	0	1	0	2	0
7	3	0	0.67	0.33	1.2	0.15
8	7	0	0.33	0.67	0.7	1
9	10	0.33	0	0.67	0.5	1
10	6	0.67	0	0.33	0.5	0.5
11	4	0.33	0.33	0.33	0.7	0.3
12	8	0.33	0.33	0.33	0.7	0.3

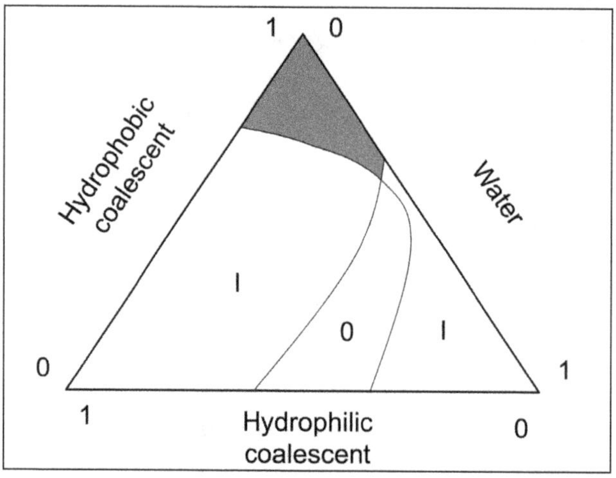

Figure 4.5: Multiple response optimization of blocking and film formation (sweet spot plot). In the grey shaded area both of the requirements are fulfilled. In region I only one of the desirability is fulfilled, in region 0 no desirability is fulfilled. These areas are generated by overlaying the contour plots for blocking and film formation.

been already analyzed in Chapter 2.6 (response surface see Figure 2.34).

Properties of the experimental design (see Table 4.3):

- Number of components: 3 (see Table 4.2)
- Number of blocks: 1
- Number of target values: 2
- Number of experiments: 12
- Randomized: yes

It is obvious from Figure 2.34 that a high content of hydrophilic solvent is negatively affecting the blocking resistance. Thus a response of blocking < 1 is only possible at low concentrations of the hydrophilic solvent over the whole concentration range of the hydrophobic solvent. Due to the fact, that the film needs a certain amount of solvent for coalescence always some hydrophobic or hydrophilic solvent is necessary (see Figure 4.4). Thus, multiple response optimization of the responses blocking and film formation is generated by overlaying the contour plots for blocking and film formation (see Figure 4.5). In two regions only one of the desirability is fulfilled, in one region no desirability is fulfilled. In one area (the grey shaded area in Figure 4.5) the complete desirability is fulfilled (commonly called **sweet spot**). Although film formation takes place without any cracks, a high content of hydrophilic solvent is impossible due to worse blocking results. Thus, only a low amount of hydrophilic, but a slightly higher amount of hydrophobic solvent is the optimized mixture to fulfil all requirements. An introduction to using trinary diagrams is given in Appendix 6.

Example: Optimization of an indoor paint

The coating formulation of an indoor paint is optimized. The dominant influence factors are the content of the fillers Calcium carbonate type 1 (C1), Calcium carbonate type 2 (C2), Kaolin (Ka), the white pigment Titanium dioxide (TiO_2) and the binder (Bi). Nine response values are analyzed: contrast ratio, viscosity, pH-value, solid content, costs, gloss 60°, gloss 85°, colour strength and rub out. It is desirable to find coating formulations, which fulfil at least 8 target values of the responses (rub out is no longer under consideration):

Table 4.3: Factor settings regarding the multiple optimization of an indoor paint

Factors	Low limit in [%]	High limit in [%]
Calcium carbonate Type 1, C1	13	19
Calcium carbonate Type 2, C2	24	34
Kaolin (Ka)	4	8
Titanium dioxide (TiO_2)	10	17
Binder (Bi)	9	13

Table 4.4: Desirability function for the multi response optimization of an indoor paint

Response	Criteria	Min	Target	Max
Contrast ratio	Target	97.5	97.8	98
Viscosity	Target	8000	11000	14000
Solid content	Target	60.5	61	62.5
Raw material costs	Minimize		0.54	0.58
Gloss 85°	Target	1	1.5	2
Color strength	Target	98.5	100	101.5

- Viscosity: 8000 - 14000 cP
- Contrast ratio: 97.5 - 98 %
- pH-value: 8.4 - 8.7
- Solid content: 60.5 - 62.5 %
- Costs of raw materials pro kg: < € 0.58
- Gloss 60°: 1.9 - 2.3
- Gloss 85°: 1 - 2
- Colour strength: 98.5 - 101.5 %

In case of 5 influence parameters and a centre point with 3 replicates we end up with 35 experiments. A fractional factorial design can reduce this amount and in addition experiment 16 was eliminated due to high viscosity afterwards.

Properties of the experimental design (see Table 4.4):

- Number of components: 5 (see Table 4.5)
- Number of blocks: 1
- Number of target values: 9
- Number of experiments: 19
- Randomized: no

Main effects and interactions are analyzed for each response. It is obvious from Figure 4.6 that the pH-value is affected by other parameters than chosen in this experimental design (e.g. amount of ammonia or sodium hydroxide). Also gloss 60° is due to the very high errors bars not significant and is neglected in the next steps.

Afterwards a desirability function is defined for the remaining 6 responses (see Table 4.4) and the overall desirability – given by the geometric average – is calculated. The sweet spot is very small (see Figure 4.7). The recommended level for the optima is at approximately 14 % Calcium carbonate type 1 (C1), 32.5 % Calcium carbonate type 2 (C2), 7 % Kaolin (Ka), 10 % binder and 11.5 % Titanium dioxide (TiO_2).

This example illustrates clearly the possibilities of DoE regarding multi response optimization: Coating formulations can be optimized according to the customer demands and in parallel a profound system knowledge is achieved.

Table 4.4: Design matrix for the multi response optimization of an indoor paint

Name of experiment	Number of experiment	Calcium carbonate 1 [%]	Kaolin [%]	TiO_2 [%]	Calcium carbonate 2 [%]	Binder [%]	Contrast ratio [%]	Viscosity [mPas]	pH-value	Solid content [%]	Cost of raw materials [€]	Gloss 60°	Gloss 85°	Color strength [%]	Rub out color strength [%]
1	12	13	4	10	24	13	93.76	4800	8	51.6	1.04	2.1	1.1	49.5	51.4
2	13	19	4	10	24	9	95.04	5560	8	55.5	0.96	2	1	74.5	73.4
3	9	13	8	10	24	9	95.97	4720	8.17	53.5	1	2.1	1.2	79.3	77.5
4	10	19	8	10	24	13	95.21	10400	8.35	61.6	1.1	2	1	63.7	67.5
5	3	13	4	17	24	9	96.97	5200	8.52	54.6	1.24	2.2	1.6	110	107.1
6	4	19	4	17	24	13	96.92	10560	8.04	62.8	1.34	2.1	1.2	101.6	100.8
7	5	13	8	17	24	13	97.29	9460	7.99	60.8	1.36	2.1	1.4	102.5	104.1
8	17	19	8	17	24	9	97.9	11700	8.05	64.6	1.3	2.1	1.4	117.8	118.1
9	6	13	4	10	34	9	96.82	7240	8.14	57.5	0.98	2.2	2.1	91.9	88.3
10	16	19	4	10	34	13	96.25	15220	8.19	65.6	1.08	2.1	1.6	77.3	77.3
11	8	13	8	10	34	13	97.04	12860	8.19	63.6	1.12	2.2	1.8	80.9	83.8
12	19	19	8	10	34	9	98.11	12220	8.25	67.5	1.04	2.2	1.9	103.5	101.8
13	14	13	4	17	34	13	97.64	14100	8.21	64.8	1.36	2.3	2.6	112.9	115.8
14	18	19	4	17	34	9	98.16	17540	8.24	68.6	1.28	2.3	2.3	126.1	127.9
15	11	13	8	17	34	9	98.97	12360	8.28	66.6	1.32	2.3	2.7	128.9	128.2
16	2	19	8	17	34	13	99.65	60000	8.08	74.8	1.42	2.3	2.6	119.8	121
17	1	16	6	13.5	29	11	97.64	9860	8.25	62.1	1.18	2.2	1.6	100	100
18	15	16	6	13.5	29	11	98.71	10700	8.32	62.1	1.18	2.2	1.8	99.9	100.2
19	7	16	6	13.5	29	11	97.2	10820	8.26	62.1	1.18	2.1	1.5	99.2	99.6

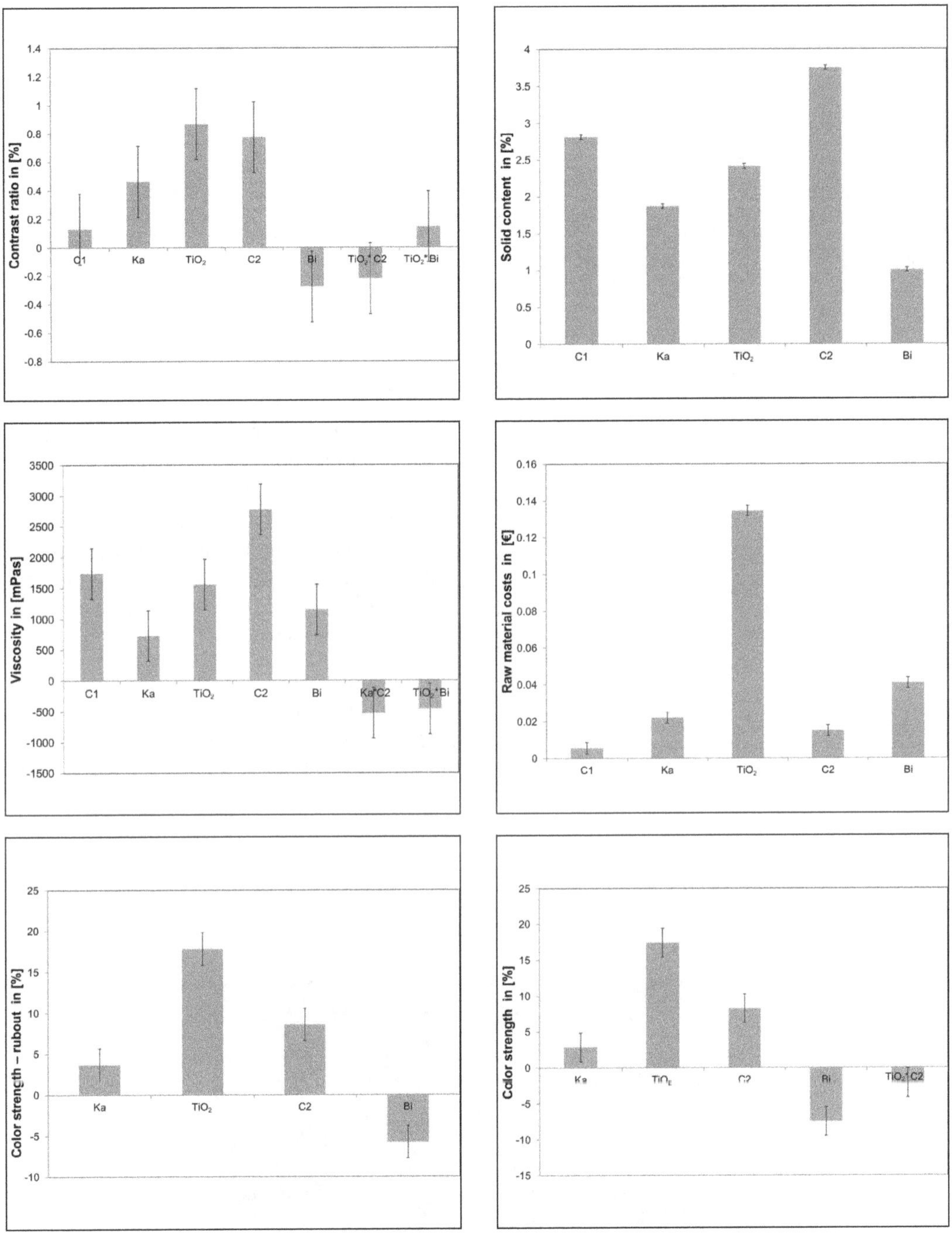

Figure 4.6: Main effects and interactions regarding the optimization of an indoor paint

4.3 Robustness or sensitivity analysis of processes and products

Chapter 4.2 emphasizes the optimization of formulations and processes regarding certain requirements. However, it is also very important to minimize simultaneoulsy the propa-

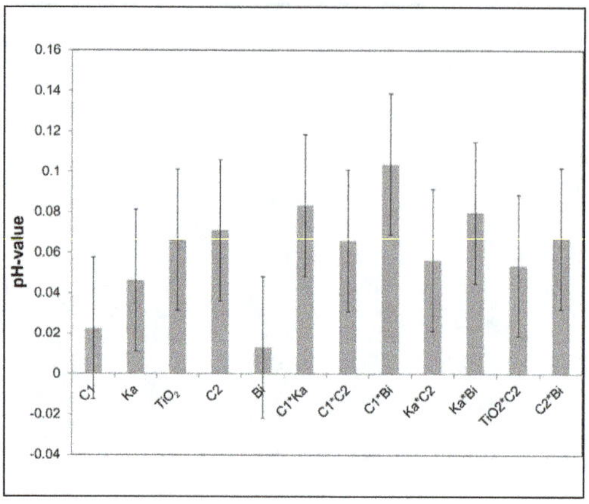

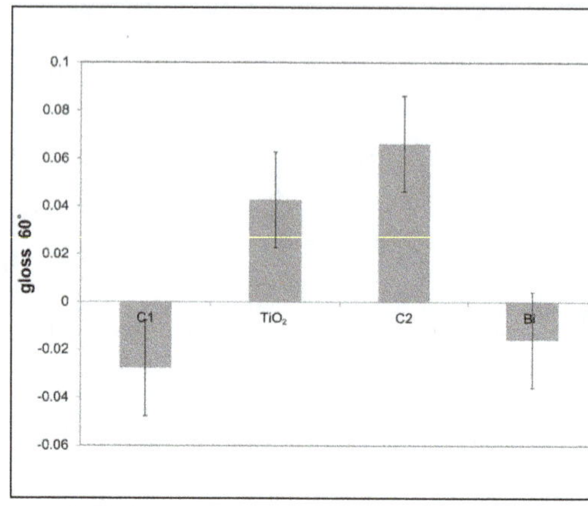

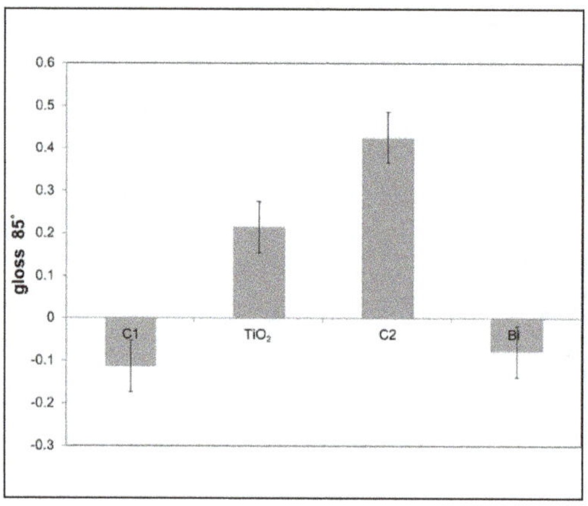

Figure 4.6 (continue): Main effects and interactions regarding the optimization of an indoor paint

gation of error to product properties due to noise. In addition, the performance should remain at a satisfying level even if some critical factors of influence (e.g. application conditions) are varying to a certain amount (**robust product design**). Variability reduction by robustness studies is one of the crucial elements of success [16].

Production process noise factors are undefined and not or hardly controllable factors of influence (e.g. quality of raw materials). In case of product application noise factors are certain application conditions which end up in larger variability (e.g. air humidity or temperature, etc.).

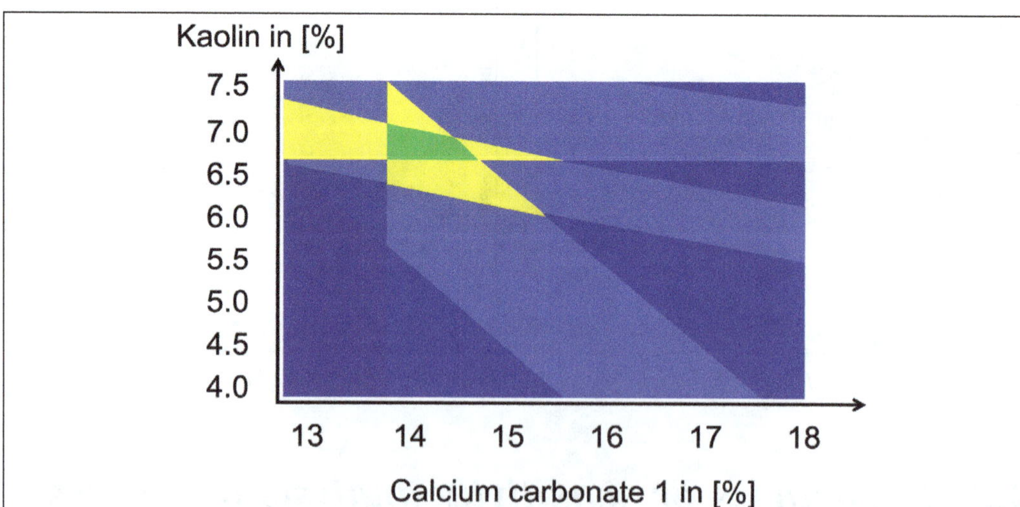

Figure 4.7: Multiple response optimization regarding an indoor paint. In the green area (sweet spot) all desired criteria, in the yellow areas 5 criteria are fulfilled. In the blue colour shade areas 4, 3 and 1 or 2 criteria are fulfilled. Increasing brightness corresponds to increasing number of fulfilled criteria. The sweet spot plot is valid for 11.5 % TiO_2, 32.5 % $CaCO_3$ (Type 2) und 10 % Binder.

For each setting inside the inner array on base of the control factors some experiments has been performed with different setting of the outer array. In this example both designs are full factorial 2^2-designs. Of course it is also possible to work with even more factors and also different numbers of levels as well as fractional factorial designs. For each of the four settings of the noise factors four data points are measured. Afterwards the mean and the standard deviation, as well as the S/N-ratio are calculated. For all three responses main effects can be calculated by the standard analysis procedure (see Chapter 2.2).

Equation 4.4: $HE(S/N) = 0.5 \cdot (a - (1) + ab - b) = 0.5 \cdot (11.93 - 15.83 + 12.39 - 24.95) = -8.23$

It is obvious from Table 4.9, that additive 1 has a large effect on the S/N-ratio. Thus, this variable is defined to minimize the variance and to optimize the robustness of the product to the environmental influences. In the third row of Table 4.9, the largest S/N-ratio of the responses is available with 24.95. Thus, the concentration of additive 1 is now adjusted to level -1, which is the relevant setting for this row (in case of more factors with a high S/N-ratio), also some parameters can be fixed). With the second variable the desired level of the number of bubbles/cm can be adjusted. Maybe other control factors without any significance are present. They should be adjusted by just optimizing the costs. Finally a proof of concept has to be done by an experiment. If the predictions are not correlating with reality, maybe interactions between control factors are present. In this situation we have to change the design. Regarding the thick-coat glaze finish an experiment is done with 0.1 % additive 1 and 1 % additive 2. Maybe the content of additive 2 has to be increased to reduce the content of micro foam even further.

Thus, by optimal design of formulation robustness to the two noise factors air temperature and humidity at the customer side can be improved. Although single formulations of the design end up with a variance of even +/- 4 bubbles/cm by variation of the noise factors, the optimal formulation has a standard deviation of +/- 0,6 bubbles/cm and, therefore, hardly any variance. Customers are able to realize between 10 °C and 30 °C as well as 50 and 80 % r.h. a constant and optimal surface quality. Be aware, that it was not necessary to limit the product specifications or to define additional quality assurance checks.

▶ ***Basic principle:*** It is not always necessary to limit the product specifications or to define additional quality assurance checks. Very often just robust formulations and application as well as production processes have to be defined to offer an adequate quality to the customer.

4.4 References

General literature references can be found in the Appendix 9.

[1] T. F. Edgar, D. M. Himmelblau, L. S. Lasdon, Optimization of chemical processes, 2nd Edition, McGrawHill, Boston, 2001

[2] U. Hoffmann, H. Hofmann, Einführung in die Optimierung: mit Anwendungsbeispielen aus dem Chemie-Ingenieur-Wesen, Verlag Chemie, Weinheim, 1971

[3] G. E. P. Box, K. B. Wilson, J. Roy. Statist. Soc. 1951, B13, 1

[4] G. E. P. Box, N. R, Draper, Das EVOP-Verfahren, Oldenbourg Verlag, München, 1975

[5] I. Gerdes, F. Klawonn, R. Kruse, Evolutionäre Algorithmen: genetische Algorithmen - Strategien und Optimierungsverfahren - Beispielanwendungen Wiesbaden, Vieweg, 2004

[6] V. Nissen, Einführung in evolutionäre Algorithmen: Optimierung nach dem Vorbild der Evolution Braunschweig, Vieweg, 1997

[7] K. Backhaus, B. Erichson, W. Plinke, R. Weiber, Multivariate Analysemethoden. Springer, Berlin, 13th Edition, 2010

[8] G. D. Rey, Karl F. Wender: Neuronale Netze. Eine Einführung in die Grundlagen, Anwendungen und Datenauswertung, Huber, Bern, 2nd Edition, 2010

[9] J. E. Vitela, E. Nahmad-Achar, R. Nakamura, A. Gama, J. Rodrı´guez, J. Coat. Technol. Res. 2007, 4 (3), 231

[10] T. Gill, J. Shutt, Scientific Computing and Automation, 1992, 5, 19

[11] R. C. Rowe, E. A. Colbourn, Chem. Educator 2003, 8, 1

[12] L. Zadeh, Information and Control 1965, 8, 338

[13] N. K. Kasabov, Neurocomputing 2001, 41, 25

[14] G. Dehringer, R. Suich, Simultaneous Optimization of Several Response Variables, Journal of Quality Technology 1980, 12(4), 214

[15] W. Kleppmann, Taschenbuch Versuchsplanung - Produkte und Prozesse optimieren, 6th Edition, Hanser, 2009

[16] M. S. Phadke, Quality Engineering Using Robust Design, Prentice Hall, London, 1990

[17] B. Gunther, A Perspective on the Taguchi Methods, Quality Progress, 1987, 44

[18] G. Taguchi, System of Experimental Design, Vol I + II, Kraus International Publications, New York, 1987

[19] T. Brinz, T. Burk, R. Emmerich, J. Ortmeier, Farbe & Lack 2009, 115(12), 32

[20] H. M. Cartwright, Application of Artificial Intelligence in Chemistry, Oxford, Oxford, 1993

[21] D. E. Goldberg, Genetic algorithms, Addison-Wesely, Reading, 1989

[22] R. Henrion, G. Henrion, Multivariante Datenanalyse, Springer Verlag, Berlin, 1995

[23] J. Zupan, J. Gasteiger, Neural networks for chemists, VCH, Weinheim, 1993

[24] F. Walters, R. Parker, S.L. Morgan, S.N. Deming, Sequential Simplex Optimization, CRC Press, London, 1991

[25] R. Schlittgen, Einführung in die Statistik, 10th Edition, Oldenbourg, München, 2003

[26] J. A. Freeman, D. M. Skapura, Neuronal Networks. Algorithms, Applications and Programming Techniques, Addison-Wesely, Reading, 1991

[27] D. Steinhausen, K. Langer, Clusteranalyse, de Gruyter, Berlin, 1977

[28] A. C. Atkinson, A. N. Donev, Optimum Experimental Designs, Oxford Press, 1992

[29] B. Klein, Versuchsplanung - DoE: Einführung in die Taguchi/Shainin Methodik, 3rd Edition, Oldenbourg Wissenschaftsverlag, München, 2011

[30] L. Eriksson, E. Johansson, N. Kettaneh-Woldm C. Wikström, S. Wold, Design of Experiments, 3rd Edition, Umetrics Academy, Umea, 2008

5 DoE-Software - do not develop the wheel once more

As already emphasized in the book, it is possible to create and analyze experimental design by hand. However, it is strongly recommended to make use of software packages from the very beginning. Otherwise it would be incredibly difficult to fit all coefficients in different models and to deal with statistical data analysis. Due to this, it is possible to concentrate on the essential aspects and to generate extensive assistance. This book gives the skills for dealing with different software packages and describes the basic definitions and terminology. The graphical and also mathematical possibilities of a computer cannot be replaced by a book.

Many of the commercial statistical software packages possess modules for design of experiments. There are also special software packages available, which assist users who are not that accomplished with the conduction of design and analysis of experiments. Naturally, this reduces the flexibility. Beside these possibilities, different simulation software packages include special design programs or modules for DoE. In principle three different types of software packages can be distinguished:

- Statistic software with tools for DoE
- Specialized software for DoE
- Simulation packages or EXCEL-based

This chapter gives a short overview without preferring a certain single solution. The examples in the book were solved partially with "MODDE" and on the other hand with "STATGRAPHICS". Both tools are common in the coating industry. „STATGRAPHICS" is mainly used in the field of automotive OEM, because it can also be used in other areas of this segment as statistic tool.

Take time for the selection. Each software solution has advantages and disadvantages. Important is the operator convenience, graphical possibilities and tools for analysis. Especially at the beginning it is not important, if a software package can perfectly deal with special plans (e.g. D-optimal). These are aspects for specialists.

Autonomous commercial software-packages for DoE:

- "DESIGN-Expert" and "DESIGN-Ease", Stat Ease Inc. (www.stat-ease.com): simple, intuitive handling. All important designs are included. Good graphical possibilities are available and, mixture designs and multi-target optimization are possible.
- "MODDE", Umetrics (www.umetrics.com): Good software for the very beginners due to self-explanatory handling. Good graphical possibilities are available and, mixture designs and multi-target optimization are possible. A demo-version can be downloaded.
- "STAVEX", Aicos Technologies (www.aicos.com): Good software for beginners, because the design is produced by asking for the demands step by step. Good screening-possibilities are available and; regarding data evaluation some requests cannot be fulfilled. A version for mixture design is available. Download of a Demo-version possible.

Statistic packages:

- “JMP”, SAS (www.jmp.com): extensive software package for statistics with an extensive package for DoE and data analysis. All relevant tools are available. Quite simple handling.
- “Minitab”, Minitan Inc. (www.minitab.com): software package for statistics with a package for DoE. Most of the designs are available and multi-target optimization is possible. Quite simple handling.
- “STATISTICA”, StatSoft (www.statsoft.de): extensive software package for statistics with modules. A package for DoE is available including all standard designs.
- “STATGRAPHICS”, StatPoint Technologies Inc. (www.statgraphics.com): extensive software package for statistics with quite simple handling. Only a few typical designs are not available, but mixture design and multi-target optimization are possible. A demo-version can be downloaded.

EXCEL-based Software:

- “EXCEL”, Microsoft: classical spreadsheet program; in this software evaluation tools for DoE are only available after self-generation (see Chapter 2.2.1). In case of data analysis (statistics) some functions are available. Especially with 3D-plots, EXCEL is very fast at the limit.
- "Sagata DoE Toolset", Sagata (www.sagata.com): spreadsheet program for DoE. A demo-version can be downloaded.

Appendix 1 - Precision, trueness and accuracy

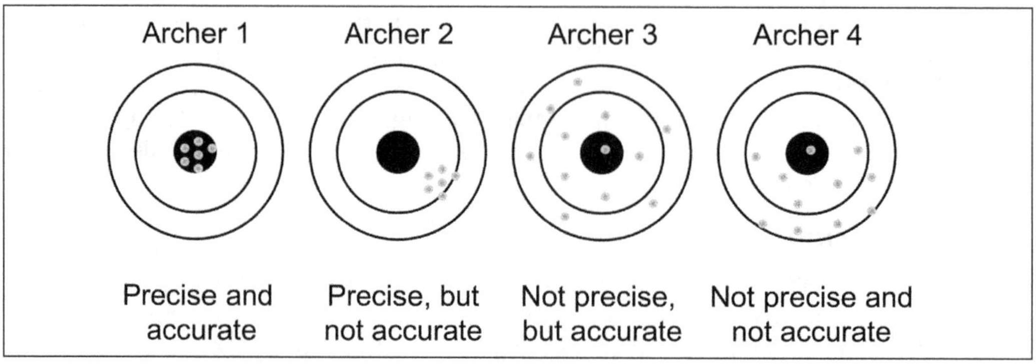

Figure A1.1: Accuracy versus precision: the target analogy in an archer contest

Let's consider an archer contest with four participants. Each one has five arrows, afterwards the results are compared (see Figure A1.1). Clear winner is archer 1. His arrows are precise in the target and he could reach the highest score. Archer 3 hit's the midpoint of the target on average, but his arrows are distributed over the whole object. So he couldn't collect enough points. Archer 4 distributed his arrows also over the whole target, but he couldn't hit the target already on average. Archer 2 could be the winner of the contest. Unfortunately he lost his glasses today, so due to his visual defect he always shot with high precision on a point below and on the right hand side of the target midpoint.

The same is true for development in coatings technology. Results should also have the highest score. Clearly, archer 1 produces the most exact result. However, archer 3 has the same result on average, although his results are scattered over a large area. Rightly, archer 1 will claim, that results from archer 3 has the same quality. Thus, in measurement technology results has to be differentiated:

- the average value of the data set agree with the true value (e.g. content of a sample) - **high trueness**. The **true value** is of theoretical nature and can never be determined directly. It's the value of a measurement which would be received in a perfect measurement without any errors. A **systematic error (bias error)** causes the mean of a data set to differ from the true value.
- single results are narrow distributed around to the average point (small spreading) - **high precision**. A **random error** causes data to be scattered more or less symmetrically around a mean value.

Both claims are summarized with the term **accuracy**, which contains of the **trueness** (affected by the systematic error) and the **precision** (affected by the random error). Thus, it's always better, to be approximately true as precise wrong.

▶ ***Basic principle:*** Every measurement is imperfect. It is impossible to perform an analysis or measurement in such a way that the results are totally free of errors or uncertainties. The true value is of theoretical nature and differs from the measured value by the systematic as well as the random error. The total deviation is described by the accuracy which consists of the trueness and the precision.

Appendix 2 - Location and spread parameters

This Appendix gives a short overview about describing a set of observations (data) as simple as possible with some parameters.

The most important parameter regarding trueness is the **arithmetic mean** (usually abbreviated to the **mean) or average value** $\bar{x}$ (Equation A2.1).

Equations A2.1: $$\bar{x} = \frac{x_1 + x_2 + \cdots + x_N}{N} = \frac{1}{N} \sum_{i=1}^{N} x_i$$

The symbol Σ is a short version for the addition of all values x_i from i = 1 to i = N. N is the total number of single results.

The **median** of a set of data is the middle result when the data are arranged by size. The arithmetic mean is highly influences from runaways. In this case it is recommended to use the median.

The **standard deviation** s is the most important parameter regarding measuring the precision. Thus, it is a way to describe the degree of agreement among several results measured in the same way. From the perspective of mathematics, it is proportional to the square root from the square sum of deviations from the mean (Equation A2.2). Very often the square of the standard deviation, the so called **variance** s^2, is used.

Equation A2.2: $$s = \sqrt{\frac{\sum_{i=1}^{N} (x_i - \bar{x})^2}{N - 1}}$$

In this formula initially the average value of the square deviation from the mean is calculated and afterwards the square root is build. If the square root is not calculated, doubling the variables $x_1, x_2, \ldots, x_N$ would build the quadruple of the average value from the square sum. If the standard deviation should vary in the same way as the data values, it is necessary to calculate the square root. Considering the formula, it becomes evident that the contribution of a single variable x_i to s is the higher the larger the distance from the average value. As the square of the values is taken, all summands are larger or at least equal to zero. Therefore, no negative contribution is formed, neither if values are above or below the average.

The term N - 1 are the so called **degrees of freedom** f. This is the number of variables, which can be chosen independent from the other values. Let's get a deeper understanding about the sense of f by looking on a mixture of 8 components. The content of each component is described by the mass fraction in %. In total only 100 % can be present. If the mass fraction of 7 components is known, the fraction of the last component can be calculated and cannot be selected independent from the other values. The degree of freedom is 7.

Equation A2.3: $$100\,\% = x_1 + x_2 + \cdots + x_8$$

In measurement technology the same principle is valid. If eight measurements are preformed and the standard deviation is calculated, it is only necessary to do once more 7 control measurements, to confirm the result, because the last measurement is anticipated by the formation of the average value.

Example: pH-value of a lime paint

The pH-value of a lime paint sample was measured several times (see Table A2.1). Calculate the average value, median and standard deviation. Derived from Table A2.1 the following statistical parameters can be calculated:

Mean: $\bar{x} = \dfrac{10.5+11.5+\cdots+13.5}{19} = 13.3$

Median: All values are arranged by size and the middle of the data set is determined.

10.5 11.5 11.5 11.5 12.5 12.5 12.5 12.5 13.5 **13.5** 13.5 13.5 13.5 13.5 14.5 14.5 15.5 15.5 16.5

Median: 13.5

Standard deviation: $s = \sqrt{\dfrac{\sum_{i=1}^{N}(x_i-\bar{x})^2}{N-1}} = \sqrt{\dfrac{43.16}{18}} = \sqrt{2.40} = 1.6$

If it is necessary to combine some standard deviations together, this can be done by the geometric mean (Equation A2.4).

Equation A2.4: $s^2 = \dfrac{s_1^2 + s_2^2}{2}$

Table A2.1: Data points and analysis by a table in case of the pH-measurement of a lime paint

x_i	$x_i - \bar{x}$	$(x_i - \bar{x})^2$
10.5	-2.79	7.78
11.5	-1.79	3.20
12.5	-0.79	0.62
11.5	-1.79	3.20
13.5	0.21	0.04
13.5	0.21	0.04
12.5	-0.79	0.62
12.5	-0.79	0.62
15.5	2.21	4.88
16.5	3.21	10.30
14.5	1.21	1.46
13.5	0.21	0.04
11.5	-1.79	3.20
14.5	1.21	1.46
13.5	0.21	0.04
15.5	2.21	4.88
13.5	0.21	0.04
13.5	0.21	0.04
12.5	-0.79	0.62
252.5		43.16

If the data set is arranged by size, a classification (ranking) can be done by determine one position on the scale before that (including the position itself), at least p % of the data (percentage based on the number of data) appear. After that point (excluding the position itself), at least (100 - p) % of the data appear. Basically these positions can be determined by counting. However, data points are very often analyzed after classification (e.g. all measurements with a pH value between 11 and 12 belong to the class pH 11 to 12).

The terminus for the value that cuts off the first n percent of the data is in statistics the so called **quantil/percentile**. In case of the 50 %-quantil, 50 % of the values are less than this value. **Quartiles** (lat. *Quartile* - quarter values) are the quantiles Q_{25} (25 %-Quantil), Q_{50} (50 %-Quantil = Median) and Q_{75} (75 %-Quantil). These three points divide the data set into four equal groups, each group comprising a quarter of the data:

- first quartile (designated Q_{25}) - 25th percentile (splits lowest 25 % of data)

- second quartile (designated Q_{50}) - median - 50th percentile (cuts the data set in two halfs)
- third quartile (designated Q_{75}) - 75th percentile (splits highest 25 % of data, or lowest 75 %)
- in the central region, between the first and the third quartile, 50 % of the data values are present.

The difference between the upper and lower quartiles is called the **interquartile range**.

The calculation of the quantil Q_i can be done for each data point very easy by determination of the rank R_i in relation to the whole set of data (see Equation A2.5). This number is available after data arrangement by size (Table 2.2).

Equation A2.5: $Q_i = \frac{R_i}{n} * 100$

with R_i equal to the so called rank of a data point and n is the total number of data points.

In case of small samples very often other calculation methods like the *Rankit* method are used (see Equation A2.6). In all such cases a certain region is eliminated at the upper and lower end of the data set and the quantils are calculated with the rest of data. This is necessary, because it is not completely secure, that the highest data point represents the 100 % of the cumulative frequency of occurrence.

Equation A2.6: $Q_i = \frac{R_i - 0{,}5}{n} * 100$

The opposite way around for each quartil Q_i a certain data point can be derived:

Equation A2.7: $R_i = \frac{Q_i * n}{100}$

If R_i is an integer, e.g. 2, 3, etc., the data point is derived by formation of the average value from the certain data point and the following point.

If R_i is not an integer, e.g. 3.12, rounding is necessary and the rounded value, e.g. 3, is used.

Table A2.2: Classified values from Table A2.1

Number of experiment	Measured value	Ranking R_i
1	10.5	1
2	11.5	2
4	11.5	3
13	11.5	4
3	12.5	5
7	12.5	6
8	12.5	7
19	12.5	8
5	13.5	9
6	13.5	10
12	13.5	11
15	13.5	12
17	13.5	13
18	13.5	14
11	14.5	15
14	14.5	16
9	15.5	17
16	15.5	18
10	16.5	19

Example: pH-value of lime paints

Arrange all values from Table A2.1 by size and calculate the quartiles Q_{25}, Q_{50}, Q_{75}.

The quartiles can be determined by Equation A2.7:

$R_i = 0.25 * 19 = 4.75$ respectively rounded: 5

The value at rank 5 is pH 12.5. Thus Q_{25} is 12.5.

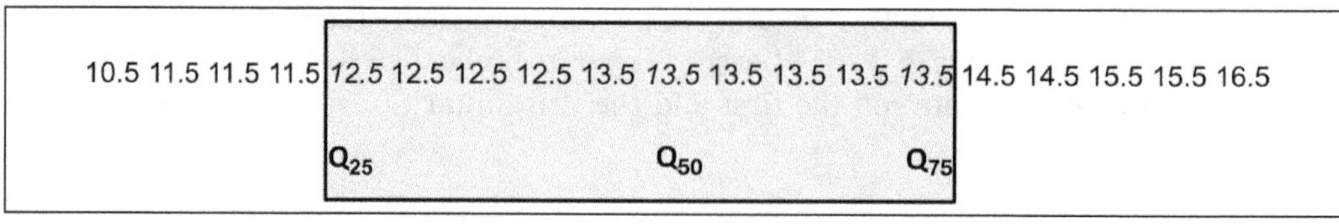

Figure A2.1: Quantile

$R_l = 0.5 * 19 = 9.50$ respectively rounded: 10

The value at rank 10 is pH 13.5. Thus Q_{50} is 13.5.

$R_l = 0.75 * 19 = 14.25$ respectively rounded: 14

The value at rank 14 is pH 13.5. Thus Q_{75} is 13.5. It is also common to use *Box* plots for the visualization. They also show the median as well as some quartils (see Figure A2.2), but different versions can be found in the literature. One very common with $Q_{2.5}$, Q_{25}, Q_{75}; $Q_{97.5}$ is used in this book.

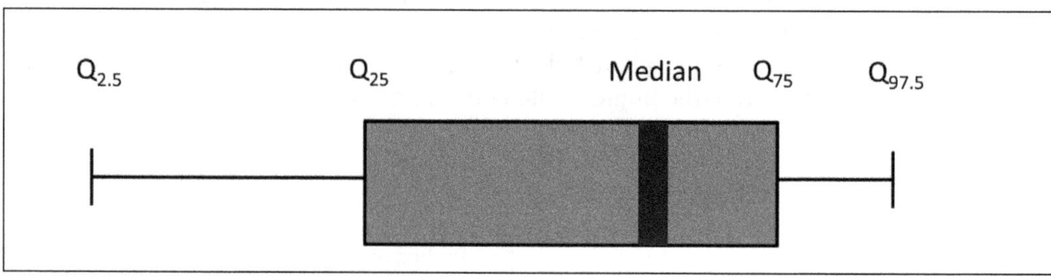

Figure A2.2: Box plot to visualize the median and the quartiles

Appendix 3 - Normal distribution

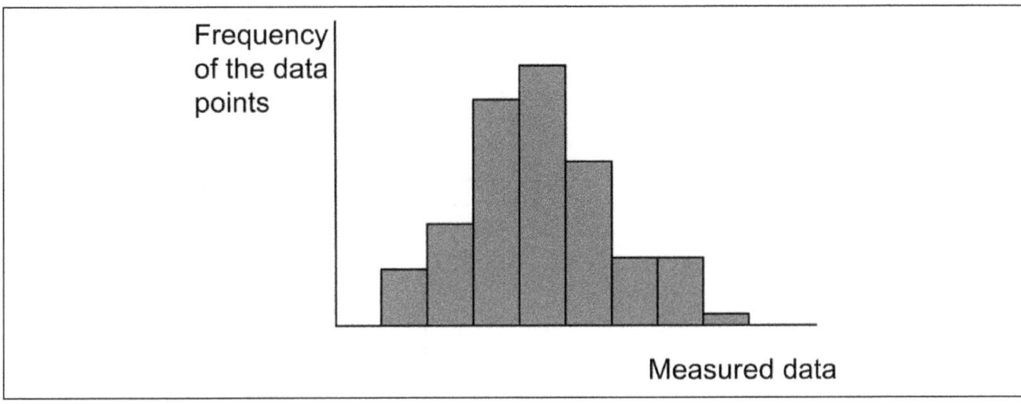

Figure A3.1: Sampling frequency distribution - histogram

Two very important definitions in statistics are the so-called population and the sample. Each single measurement is individual. Think about the noise factors like actual air pressure, vibrations of the floor, etc. Thus it is not enough to analyze just single experiments. On the other hand it is impossible to do an infinite number of measurements due to lack of time and cost aspects. As alternative usually a certain small representative fraction of the population is analyzed, the so called **sample**. The set of all in principle possible measurements and experiments is called **population**.

Example: All batches of a produced coating is the population. The sample is just the small number of batches, which were really analyzed. Maybe only 10 batches of totally 100 are analyzed.

▶ ***Basic principle:*** Population covers the whole group about which a statistical statement should be made. The sample is a random selection out of the population. As long as this selection is representative, it is possible to predict from the sample some properties of the population.

Very often general properties and laws can be observed, if many samples are collected and analyzed. This is especially the case with the frequency of appearance of measured values versus the measured values itself (see the **histogram** in Figure A3.1). If only a small number of data is analyzed, single data points occur. With increasing number of data, especially in case of an infinite number, a continuous curve is generated, because some results appear more often than others. In case of physical and chemical measurements very often a so called **Gaussian normal distribution** with a typical bell-shaped curve around the mean occurs.

With an increasing number of sample data, the sample mean $\bar{x}$ comes closer to the population mean. In case of no systematic error, the population mean is equal to the **true value** μ **(population mean)**. In addition, it has to be distinguished between the sample standard deviation s and the **population standard deviation** σ. Mean and standard deviation s are good estimations of the parameters of the population in case of enough data. However, when N is small, $\bar{x}$ differs from μ, because a small sample of data does not exactly represent its population. Thus, in case of a tiny sample it is much better to use the robust median as

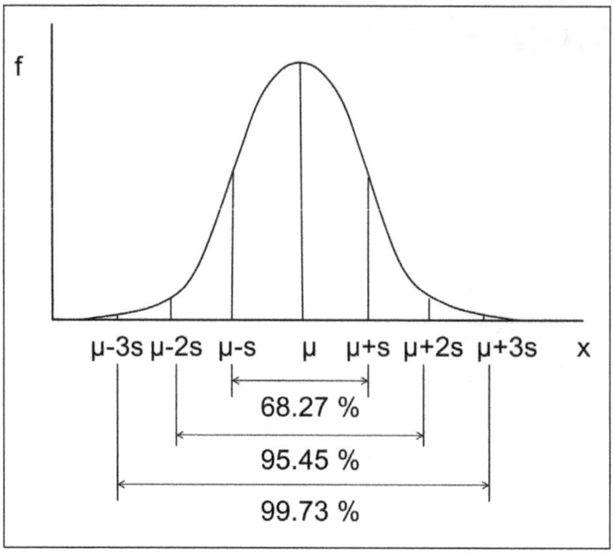

Figure A3.2: Gaussian normal distribution error curves and sense of the standard deviation

estimation of μ. To account for the variability of the standard deviation s, the so called **student factor** is used (see Appendix 4).

Three properties of the Gaussian curve are important:

- The curve is symmetric, positive and negative errors has the same frequency
- The relative frequency of data with small deviation is large. 68.27 % of the data are between, in the region +s and -s, so ± 1 s around the mean μ.
- 95.45 % of the data are between, in the region ± 2 s around the mean μ.
- The relative frequency of data with large deviation is small. 99.73 % of all data are in the region ±3 s, so only 0.27 % are outside this area.

Very often it is necessary to deal with questions regarding the number of data below or above a certain value – e.g. the number of data below a certain pH-value. This can be consi-

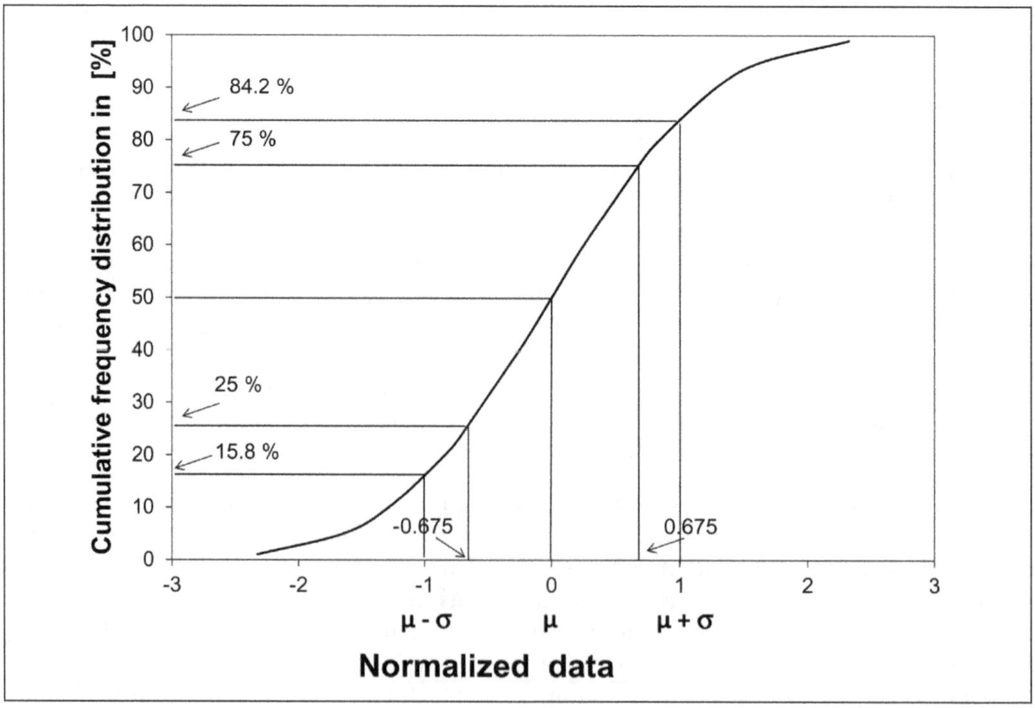

Figure A3.3 Quantiles of the normal distribution

Appendix 5 - Hypothesis, tests and conclusions - statistical tests

Until now only samples were analyzed and statistical parameters calculated. However, this never gives a certain validity regarding observations. Statistical tests support in these situations and are sharpening judgments. In order to explain the observation, a hypothetical model is advanced and the validity tested.

Let's look on an example: The scratch resistance of a certain coating (in Newton) is under examination. The coating manufacturer defined in his specifications an average value of $\mu = \mu_0$ and this should be checked based on a certain sample with a useful statistical tool. First of all, we define a so-called primary or **null hypothesis** with the statement of the manufacturer.

$H_0 : \mu = \mu_0$.

With other words, we postulate that the two quantities are the same and differences are attributed to random errors. As alternative we define the **alternative hypothesis**:

H_A: $\mu \neq \mu_0$

This hypothesis postulate, that the average scratch resistance μ is different from μ_0. So, if we can conclude at the end of the test, that the specification of the manufacturer is false, we can assume that the alternative hypothesis H_A is true.

To check this, n batches of the coating were selected and the scratch resistance determined. Afterwards, the mean of the sample $\bar{x}$ is compared with the specification μ_0. Now, it is necessary to find a parameter, which can define, whether the observed difference is significant or not. Usually, if the observed difference is greater than or equal to the difference that would occur 5 times in 100 (the 5 % probability level), the null hypothesis is considered to be questionable and the difference is judged to be significant. Still, there is a risk of 5 % that the difference arises solely as a result of random errors. Thus, this **level of significance** α should be small.

Usually a test parameter T is calculated based on the question. In this example, the test parameter is already the average value. Test variable T and the significance level α determine a certain critical value PG and a **critical interval**, defined by

$\mu_0 - PG \leq \bar{x} \leq \mu_0 + PG$

is calculated.

Thus, we can judge whether a difference is present or not. Now, two cases have to be distinguished:

- If the mean of the sample $\bar{x}$ is inside this interval, the deviation between $\bar{x}$ and μ is only random based. The null hypothesis is retained and no evidence of a deviation can be determined. The specification is okay.
- If $\bar{x}$ exceeds the critical interval, the null hypothesis is rejected. The probability, that the difference between $\bar{x}$ and μ_0 is only random, is with α very small. Thus, the null hypothesis is not tenable and the specification is wrong.

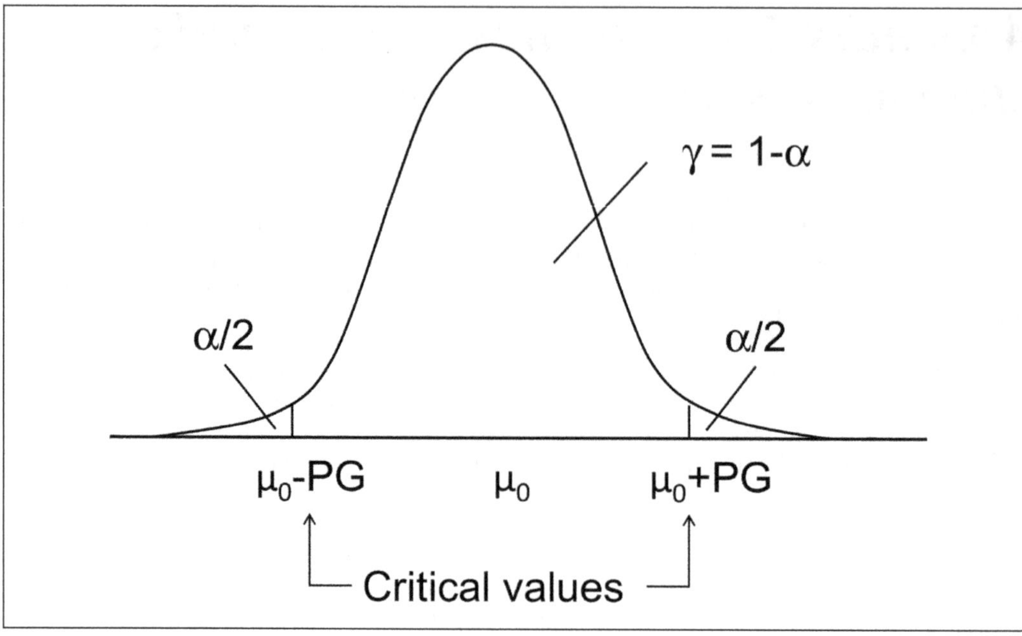

Figure A5.1: Calculation of the critical limits

At the end of each test a decision has to be judged. This can be for the benefit of the null hypothesis or the alternative hypothesis. However, we have to take into account always, that there is not 100 % security. In case of such test, there is always a certain probability of retaining a null hypothesis when it is in fact false. In principle two types of error are distinguished:

- An **error of type 1** is present, when the null hypothesis is rejected even though it is true. In other words, we think that there is an effect, although it's not true. The probability, or the small risk for such an error is α and equal to the level of significance.
- An **error of type 2** is present, when the null hypothesis is retained even though it is false. In other words, we think that there is no effect, although there is one in practice. The probability, or the risk, for such an error is β. Apart from this term very often the so-called power of a test is used and calculated by $1 - \beta$. It's the probability of rejecting the null hypothesis when it is false and therefore the probability of making a correct decision.

In both cases a wrong decision takes place and α and β are the risk of accepting false hypothesis. In addition, every reduction of the type 1 error increases the type 2 error and vice versa. Thus, it has to be decided from case to case, which error is more important. This is in contrast to the main requirement of statistical tests, because they should prevent from wrong conclusions. Certainly there is some protection, but it's limited with a certain risk to come to a wrong decision. Statistics cannot make absolute statements (see Appendix 4). In case of 19 from 20 experiments the true value will be inside the 95 % confidence interval and we will derive the right conclusions. In case of the remaining experiment a significant different value is generated and a wrong statement will be given. If we would like to limit this risk, we can deal with 99 % confidence intervals, but we must anticipate still 1 of 100 experiments with a wrong-positive conclusion.

Example: Picking mushrooms

Let's look on a person picking up mushrooms and deciding if a mushroom is toxic or not. Which errors should be small in case of the null hypothesis "the mushroom is toxic"?

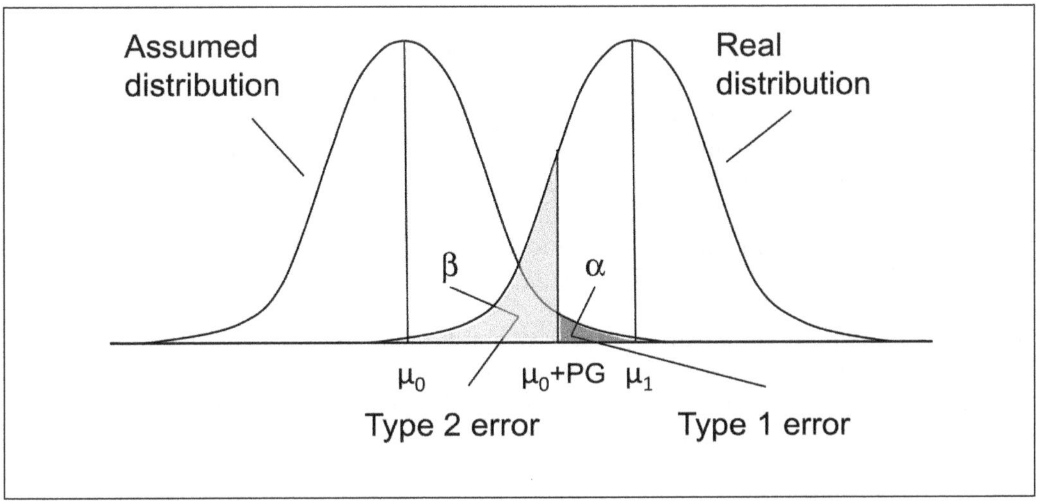

Figure A5.2: Type 1st and 2nd errors

In practice the α-level is decided first of all. Afterwards the critical value is calculated and the probability for the error of type 2 as a function of the position of μ.

From a perspective of statistics, the judgement about a hypothesis is choosing a critical value on a one-dimensional scale, it must be decided between two cases:

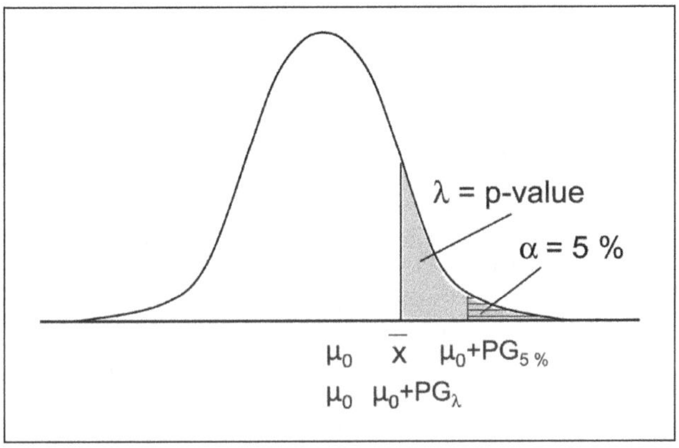

Figure A5.3: p-value - the propability, with which the mean is just equal to the critical value PG, is described by the gray shaded area (one-tailed test).

- **One-tailed (or sided) test**: questions if a property is above or below a certain value.
- **Two-tailed (or sided) test**: question, if a property is between two limits. Thus, it is necessary to define two limits and to define the propability that the value is inside the interval or outside (Figure A5.1). This is approach is typically used in case of confidence intervals (see Appendix 4).

p-value

The significance level α (e.g. 5 %) has to be decided always in case of a test. Afterwards the critical values $PG_{5\%}$ can be calculated and a judgment has to be done. However, it is also possible to go the opposite direction and to calculate the probability, which reject the null hypothesis just by the test variable. Thus, starting from the test variable $\bar{x}$ of the sample, the probability α is calculated, which corresponds to a critical value based on this significance level $\mu_0 + PG_\alpha$ is exactly equal to $\bar{x}$ (one-tailed test). This probability is called **p-value** (p for probability) (see Figure A5.3).

This parameter indicates, if a test value is inside or outside the critical interval. In addition, also information about the distance from the critical limit is present. The illustration in Figure A5.3 helps deriving the following correlations between the calculated area and the test result (α = 5 %), one-tailed test:

- Area (p-value) < 0.001: measured value of the sample inside the critical interval $x > \mu_0 + PG$. Thus a highly significant effect is present.
- Area (p-value) < 0.01: measured value of the sample inside the critical interval $x > \mu_0 + PG$. Thus a very significant effect is present.
- Area (p-value) < 0.05: measured value of the sample inside the critical interval $x > \mu_0 + PG$. Thus a significant effect is present.
- Area (p-value)= 0.05: measured value of the sample is equal to the critical interval $x = \mu_0 + PG$.
- Area (p-value) > 0.05: measured value of the sample inside the critical interval $x < \mu_0 + PG$. Thus no significant effect is present.

▶ ***Basic principle:*** The p-value is the probability, that we think there is an effect although it's not the case in reality. In case of a 95 % confidence interval effects are significant when p is smaller than 0.05. Both considerations are equal and two definitions of significance do not exist.

One test, the so-called **t-test** is also described also in Appendix 4 and Chapter 1.6. The calculation of the confidence limit enables also the check, if two means 1 and 2 are different. As test variable the value $t = \frac{\bar{x} - \mu_0}{s/\sqrt{N}}$ is used. In this equation, t is the already described student factor

Table A5.1: Values of F for a probability of 95 % and different degrees of freedom

f_2 \ f_1	1	2	3	4	5	10	20	50	∞
1	161	199	216	225	230	242	248	252	254
2	18.5	19.0	19.2	19.2	19.3	19.4	19.4	19.5	19.5
3	10.1	9.55	9.28	9.12	9.01	8.79	8.66	8.58	8.53
4	7.71	6.94	6.59	6.39	6.26	5.96	5.80	5.70	5.63
5	6.61	5.79	5.41	5.19	5.05	4.74	4.56	4.44	4.37
10	4.96	4.10	3.71	3.48	3.33	2.98	2.77	2.64	2.54
20	4.35	3.49	3.10	2.87	2.71	2.35	2.12	1.97	1.84
50	4.03	3.18	2.79	2.56	2.40	2.03	1.78	1.60	1.44
∞	3.84	3.00	2.60	2.37	2.21	1.83	1.57	1.35	1.00

Table A5.2: Values of F for a probability of 99 % and different degrees of freedom

f_2 \ f_1	1	2	3	4	5	10	20	50	∞
1	4052	4999	5404	5624	5764	6056	6209	6302	6366
2	98.5	99.0	99.2	99.3	99.3	99.4	99.4	99.5	99.5
3	34.1	30.8	29.5	28.7	28.2	27.2	26.7	26.4	26.1
4	21.2	18.0	16.7	16.0	15.5	14.5	14.0	13.7	13.5
5	16.3	13.3	12.1	11.4	11.0	10.1	9.55	9.24	9.02
10	10.0	7.56	6.55	5.99	5.64	4.85	4.41	4.12	3.91
20	8.10	5.85	4.94	4.43	4.10	3.37	2.94	2.64	2.42
50	7.17	5.06	4.20	3.72	3.41	2.70	2.27	1.95	1.68
∞	6.63	4.61	3.78	3.32	3.02	2.32	1.88	1.52	1.00

(Appendix 4) and s is the geometric mean of the standard deviation of the two compared means. Thus, two means must have at least a difference of $2 * \frac{t * s}{\sqrt{N}}$ to declare a significant difference (α and β error are equal) (see Figure 1.22).

Another important test is the so-called F-test (see also Chapter 3.4), which compares standard deviations, i.e. the random errors of two sets of data. It has to be decided, if the difference is just based on the random error (meaning, that both estimations of the variance s_1 and s_2 can be attributed to the same standard deviation of a normal distributed population of data), or not. To compare two variances s_1^2 and s_2^2, the F-test considers the ratio of the two sample variances. The quantity F defined by s_1^2/s_2^2 is the test variable and should follow a F-distribution, if the null hypothesis $\sigma_1^2 = \sigma_2^2 = \sigma^2$ is true. The hypothesis has to be rejected if $F > F(P, f_1, f_2)$. Thus, between the two variances s_1^2 and s_2^2 is a significant difference and $\sigma_1^2 > \sigma_2^2$.

The calculation of the corresponding p-value is not very simple. Depending on the degree of freedom, it is necessary to collect the value of the test variable in the statistical tables of the F-distribution. The significance level of the table, which expresses the test variable at the present degree of freedom, represents the p-value. In practice this approach is not possible, but simulation tools are available.

Example: Comparison of two standard deviations

Comparison of two standard deviations with 1 and 3 degrees of freedom (f_1= 1, f_2= 3). The test variable is 10.1. Calculate the p-value.

The value of the test variable is exactly equal to the number in Table A5.1. Thus, the p-value is 0.05 and the probability, that the ration of both standard deviations is 10.1 even though both variances differ from each other, is 5 % (α-error).

Example: ANOVA - comparison of two square sums

Comparison of two square sums during an ANOVA.

SQ_1 (f = 1) = 0.056

SQ_2 (f = 3) = 0.004

The ratio from SQ_1/SQ_2 is 12.56. The degrees of freedom have to be taken into account, because only standard deviations and not only square sums can be compared. Thus, the ration has to be multiplied by 3, because SQ_2 has to be divided by 3 (see also Chapter 3.5). This ends up in a ratio of 37.935.

It is possible to determine from Table A5.1 an F-value regarding 5 % probability and f_1 = 1 and f_2 = 3 with F(95 %, 1, 3) = 10.10. The test variable is with 37.935 much higher than the critical value F(95 %, 1, 3). Thus, these two square sums have not the same distribution. Also for a probability of 1 % (see Table A5.2), the value F(99 %, 1, 3) = 34.1 is smaller. Thus, the corresponding p-value for the test variable 37.935 is < 0.01.

References

General literature references can be found in the Appendix 9.

- L. Papula, Mathematik für Ingenieure und Naturwissenschaftler, Bd. 3, Vieweg Verlag, 1994
- W. A. Stahel, Statistische Datenanalyse, Vieweg Verlag, 1995
- E. Kreyszig, Statistische Methoden und ihre Anwendungen, Vandenhoeck & Ruprecht, Göttingen, 4. Auflage, 1973

Appendix 6 – Three-component diagrams

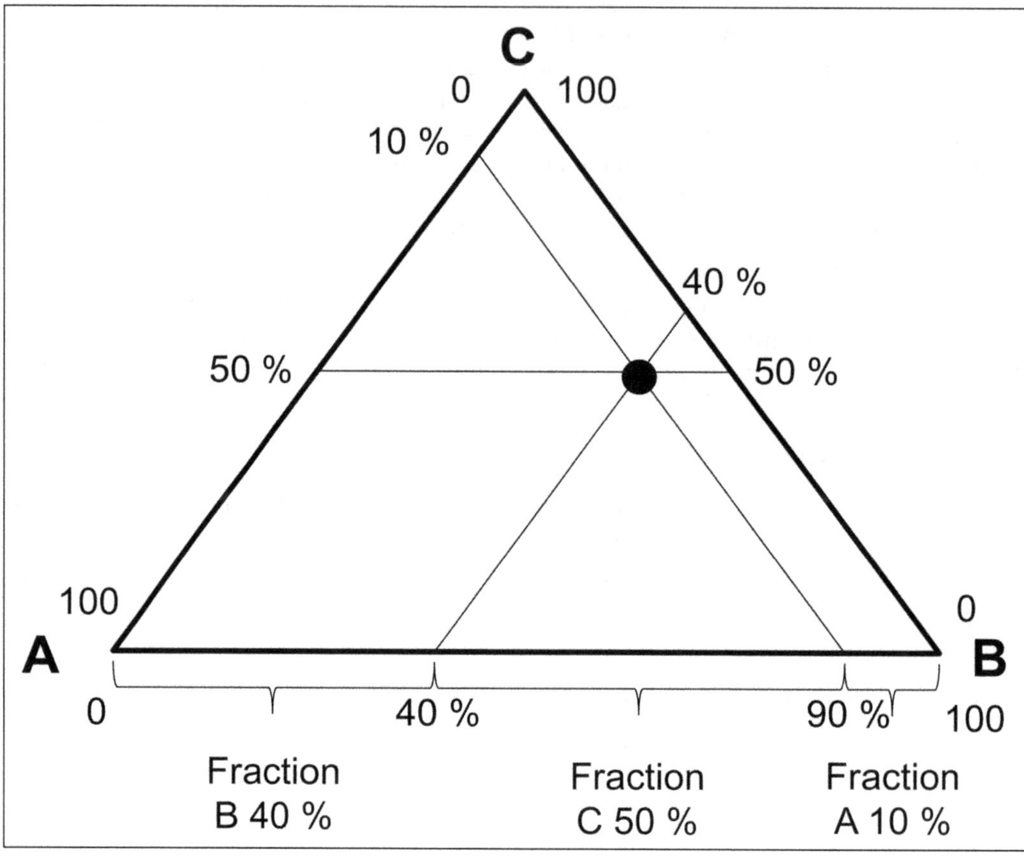

Figure A6.1: Interpretation of Rozebum's three-component diagrams

Appendix 7 - Linear regression

In this appendix the calculation of linear regression is described. For details it's referred to special literature.

Basic Equation (Equation 1.9):

Equation A7.1: $y = a_0 + a_1 * x$

Slope a_1:

Equation A7.2: $a_1 = \dfrac{\sum[(x_i - \bar{x}) * (y_i - \bar{y})]}{\sum(x_i - \bar{x})^2}$

Intercept a_0:

Equation A7.3: $a_0 = \bar{y} - a_1 * \bar{x}$

with

Equation A7.4: $\bar{x} = \dfrac{1}{N} * \sum x_i$

Equation A7.5: $\bar{y} = \dfrac{1}{N} * \sum y_i$

with N as the total number experiments

In addition:

Equation A7.6: $Q_{xx} = \sum x_i^2 - \left(\left(\sum x_i\right)^2 / N\right)$

Equation A7.7: $s_{xo} = \dfrac{\sqrt{\dfrac{\sum(y_i - \hat{y}_i)^2}{N_K - 2}}}{a_1}$ *with* $\hat{y} = a_0 + a_1 * x$

The approach to define the result for a data point $\hat{y}$ is:

Equation A7.8:

$$\hat{x} \pm VB\ (\hat{x}) = \frac{\hat{y} - a_0}{a_1} \pm s_{xo} * t(95\ \%\ f = N_K - 2) * \sqrt{\frac{1}{N_K} + \frac{1}{N_a} + \frac{(\hat{y} - \bar{y})^2}{a_1^2 * Q_x}}$$

with

$\hat{y}$ average value from N_a measurements (it is also possible that $N_a = 1$, meaning a single data point without any repeat)

N_K = number of data points for the regression curve

Example: Estimation of the glass transition temperature via DSC - Chapter 3.3.4

With the above mentioned equations the data from Chapter 3.3.4 - Example: Estimation of the glass transition temperature via DSC has been calculated.

Table A7.1 Calculation values for the regression analysis of the example from Chapter 3.3.4

	x_i	y_i	$x_i - \bar{x}$	$y_i - \bar{y}$	$(x_i - \bar{x})^2$	x_i^2	$\hat{y}_i$	$(y_i - \hat{y}_i)$	$(y_i - \hat{y}_i)^2$
	10	40	-10	-20.54	100	100	39.7	0.3	0.09
	15	52	-5	-8.54	25	225	49.7	2.3	5.29
	20	57.3	0	-3.24	0	400	59.7	-2.4	5.76
	25	72.7	5	12.16	25	625	69.7	3	9
	30	80.7	10	20.16	100	900	79.7	1	1
Sum	100	302.7	0	0	250	2250			21.14
Average	20	60.5							

References

- W. Funk, V. Dammann, G. Donnevert, Qualitätssicherung in der Analytischen Chemie, 2. Edition, Wiley-VCH, Weinheim, 2005

Appendix 8 – Failure mode and effect analysis, FMEA

Failure Mode and Effect Analysis (FMEA) is a method, which eliminates or reduce failures starting the ones of highest priority. Thus, FMEA is used in quality management to minimize the risk due to the appearance of errors. Evaluation and quantification of risk areas is in focus as well as defining possibilities to decrease their impact as soon as possible. The goal is to find potential errors of a product or process already during the planning step. Thus, FMEA is a type of preventive error prevention instead of an identification and correction afterwards. Mistakes which are not made, has not to be corrected!

In a first step (system analysis) the system and all included processes are described and defined. Single system elements are defined by an arrangement and determination of the interfaces between the elements. A form assists during this step (see Table A8.1). During the following failure analysis each possible failure is assigned to certain elements. In this step it is important to look on all possible types of failures without taking into account the probability of occurrence. Each identified failure is analyzed regarding the consequences on the system, e.g. on the user of the product. Each point is a row in the form, because each potential failure is combined with a potential failure consequence and reason. In the last step of the analysis all possible reasons, which will end up in a certain errors are documented. Afterwards possibilities to avoid the failure and possibilities to enhance the detectability are listed.

In the following risk analysis the probabilities for **Occurrence (O)**, **Severity (S)** and **Detectability (D)** of a failure are evaluated. The evaluation is performed by the so-called risk priority number (RPN), which is calculated by:

Equation A8.1 Risk priority number (RPN) $= S \cdot O \cdot D$

The scale of each parameter is running from 1 to 10 with 10 being worst. RPN ranges between 1 and 1000. If RPN is above 125 improvement measures should be done. In this case, avoiding of failures has always a higher priority than an improved rate of detection.

Table A8.1: FMEA form

Component	Type of failure	Effect	Reasons	Present situation					Improvement	Responsible	Improved situation				
				Control	S	O	D	RPN			Effect	S	O	D	RPN
Pump	Pump is not working	A component cannot react	Forget to switch on	Operating instruction	8	3	3	72							
			Power cut	No	10	1	1	10							
			Membrane defect	No	10	2	3	60	Flow control		Enhanced detectability	10	2	2	40
S = Severity				O= Occurence							D = Detectability				

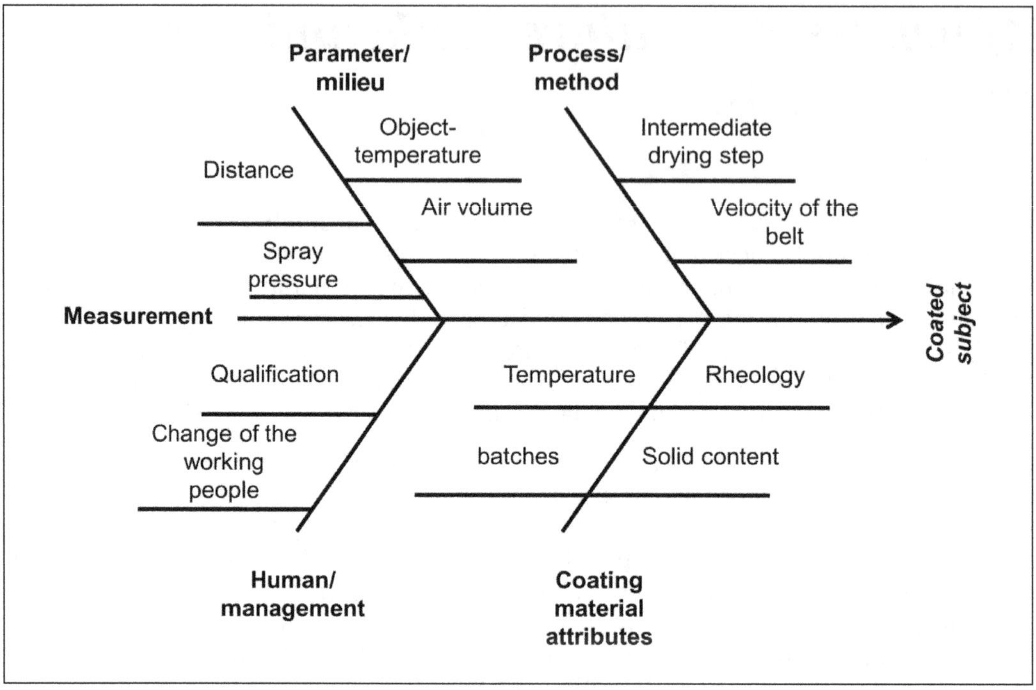

Figure A8.1 Ishikawa-Diagram

The sequence of optimization should be done based on the *Pareto* principle: generation of 80 % success by solving 20 % of the problems. With the priority sequence based on FMEA available resources can be used in an optimized way. Thus, FMEA perfectly suits as partner for DoE, because all relevant parameters of a system are selected.

Influencing factors can be structured with a cause and effect diagram (so called ***Ishikawa*-diagram or fishbone-diagram**, see Figure A8.1). In this diagram all possible reasons for a certain problem are illustrated by main loads and branches based on the so called **7M approach**, drawn from total quality management. Usually the following main reasons: human, material, method, measurement, machine, environment are analyzed.

The Ishikawa-diagram is generated always in the same way:

- Description and identification of the problem
- Selection of the main reasons and noting on the main load in the diagram
- Selection of the reasons with second priority and noting them on the branches in the diagram

This diagram is the perfect base for an FMEA, because many reasons for a problem are illustrated. Thus, a structured and organized procedure is possible. This tool is perfectly suited for team work.

References

- W. Eversheim, Qualitätsmanagement für Dienstleister, Verlag Springer, Berlin, 2000
- G.F. Kamiske, J-P. Brauer, Qualitätsmanagement von A bis Z, Verlag Hanser München, 2006
- T. Tietjen, D.H. Müller, FMEA Praxis, Verlag Hanser München, 2003
- S. Kromidas, Qualität im analytischen Labor, VCH, Weinheim, 1995
- T. Pfeifer, Qualitätsmanagement, Hanser Verlag, München, Wien, 2. Auflage, 1996

Appendix 9 - General references

- D. C. Montgomery, Design and Analysis of Experiments, 7. Edition, Wiley & Sons, 2008
- Z. R. Lazic, Design of Experiments in Chemical Engineering, Wiley-VCH, Weinheim, 2004
- G. E. P. Box, J. S. Hunter, William G. Hunter, Statistics for Experimenters: Design, Innovation and Discovery, 2. Edition, Wiley & Sons, 2005
- Dean, D. Voss, Design and Analysis of Experiments, 2. Edition, Springer, Berlin, 2003
- R. H. Myers, D. C. Montgomery, C. M. Anderson-Cook, Response Surface Methodology: Process and Product Optimization Using Designed Experiments, 3. Edition, Wiley & Sons, 2009
- G. M. Cox, William Cochran Experimental Designs, Wiley 1950, 2. Edition, 1996
- Orth, S. Soravia, Design of Experiments, in: Ullmann's Encyclopedia of Industrial Chemistry, 6th Edition, Wiley-VCH, Weinheim, 2003
- G. E. P. Box, N. R. Draper, Empirical Model-Building and Response Surfaces, John Wiley & Sons, New York, 1986
- E. Morgan, Chemometrics: Experimental Design, John Wiley & Sons, Chichester, New York, 1995
- S.N. Deming, S.L. Morgan, Experimental Design, Elsevier, Amsterdam, 1993
- C.K. Bayne, I.B. Rubin, Practical experimental design and optimization methods for chemists, VCH, Weinheim , 1986
- J.A. Rice, Mathematical Statistics and Data Analysis, 3. Edition, Duxbury Press, Belmont California, 2007
- D.K. Ludlow, K.H. Schulz, J. Erjavec, J. Eng. Educ. 1995, 84, 351

Acknowledgements

The author would like to thank everybody who has assisted with the preparation of this book. Many people were involved already in the German version, which has been published in 2011. I am grateful to Sabine Ansorge, Wörwag, Stuttgart, Germany for so many helpful advices and most of the examples in the book. This adds a strong application flavor to the book and makes it useful as a reference tool. Acknowledgement to Markus Fessler, Wilfrid Jochum, Rudolf Kamaun, Simon Lackner and Egon Rubatscher ADLER-Werk Lackfabrik, Schwaz, Austria for their help in the preparation of some examples and the fruitful discussions. Jochen Blumberger, Department of Physics and Astronomy, University College London, thank you very much indeed for your assistance during the translations of some parts. Many thanks to the executive board and the owners of ADLER-Werk Lackfabrik, Schwaz, Austria for the possibility to establish this method in the company as well as the support and the confidence in the last years.

Lust but not least I would like to thank my family, especially my wife, who had so much understanding and sympathy to this hobby. In addition, thank you very much for the assistance during the proof reading. Without her support, patience and unwavering love this book would not have been possible.

Innsbruck, Mai 2014

Albert Rössler

www.ingramcontent.com/pod-product-compliance
Lightning Source LLC
LaVergne TN
LVHW081519300926
849161LV00006B/1222

* 9 7 8 3 8 6 6 3 0 8 8 5 5 *

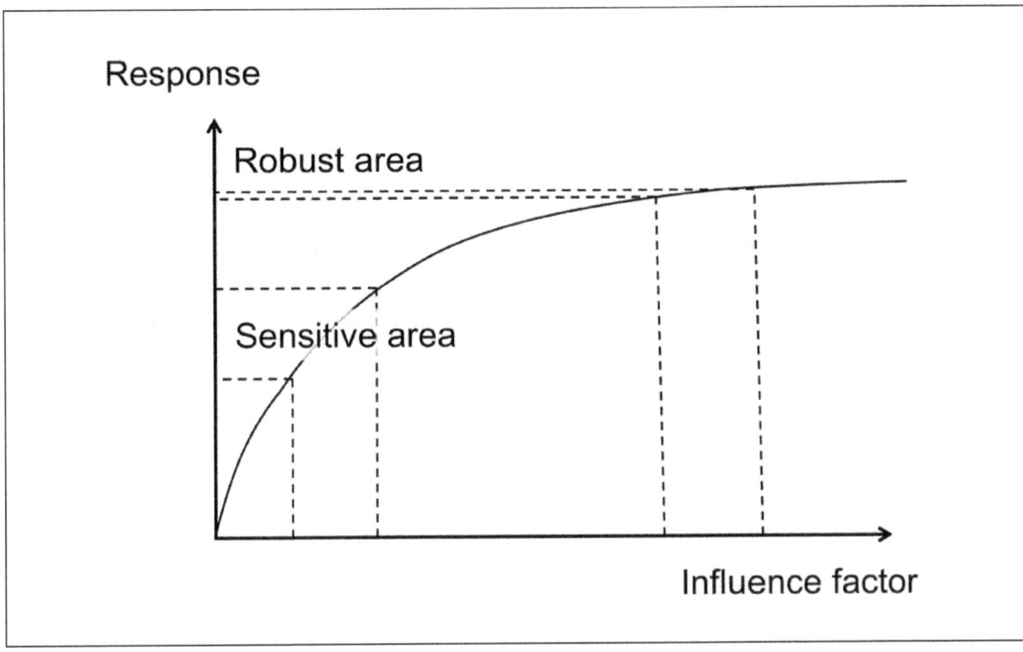

Figure 4.8: Propagation of error in case of non-linear effects

▶ ***Basic principle:*** It is very important to minimize simultaneously to the optimization the sensitivity to varying inputs or noise in systems, processes or products (robust product design). Variability reduction by robustness studies (sensitivity analysis) already during the development of a product is one of the crucial elements of success.

How shall we adjust our factors to guarantee robustness? DoE offers some helpful methods to minimize the propagation of error:

- Qualitative analysis of the response surface
- Quantitative analysis of the response surface
- *Taguchi*-method

The first two methods are part of the response surface method (RSM).

4.3.1 Qualitative analysis of the response surface

The RSM method produces maps of the product performance or the system behaviour, similar to topographical displays of elevation, as a function of the controllable input variables. The objective is now to find the plateaus of product quality and process efficiency which are not sensitive to variations in component levels or certain critical factor settings. You can find these desirable operating regions visually just by looking over the 3-D renderings of response surfaces.

Example: Disturbance in levelling of a pigmented base coat

Customers claim consistently the levelling of a pigmented base coat (analyzed by the long wave – LW). To improve this response value and to find a robust process window, the effect of binder and solvent has been studied in 5 levels in both, the old and a new rheology module [19].

Table 4.6: Factor settings regarding the rheology optimization of a base coat

Factors	Low level in [%]	High level in [%]	steps
Solvent	22.0	28.0	5
Binder	25.0	35.0	5

Table 4.7: Experimental design regarding the rheology optimization of a base coat

Number	Factors		response	
	Solvent in [%]	Binder in [%]	LW [10–15 µm] old rheology modul	LW [10–15 µm] new rheology modul
1	22	25	8.6	7.3
2	23.5	25	4.7	9.2
3	25	25	3	8.3
4	26.5	25	2.8	8.2
5	28	25	1.5	11.3
6	22	27.5	11.5	3.0
7	23.5	27.5	10.4	3.2
8	25	27.5	6.1	3.6
9	26.5	27.5	3.2	3.0
10	28	27.5	2.5	3.9
11	22	30	11.5	3.2
12	23.5	30	10.8	2.7
13	25	30	6.4	2.7
14	26.5	30	4.5	2.2
15	28	30	2.6	3.1
16	22	32.5	12.1	2.5
17	23.5	32.5	10.3	2.1
18	25	32.5	6.7	3.3
19	26.5	32.5	4.8	2.4
20	28	32.5	3.2	1.8
21	22	35	11.9	2.5
22	23.5	35	10.2	3.4
23	25	35	8.8	2.2
24	26.5	35	7.1	4.7
25	28	35	3.4	1.1

Properties of the mixture design (see Table 4.7):

- Number of components: 2
  (see Table 4.6)
- Number of blocks: 1
- Number of target values: 1
- Number of experiments: 25
- Randomized: no

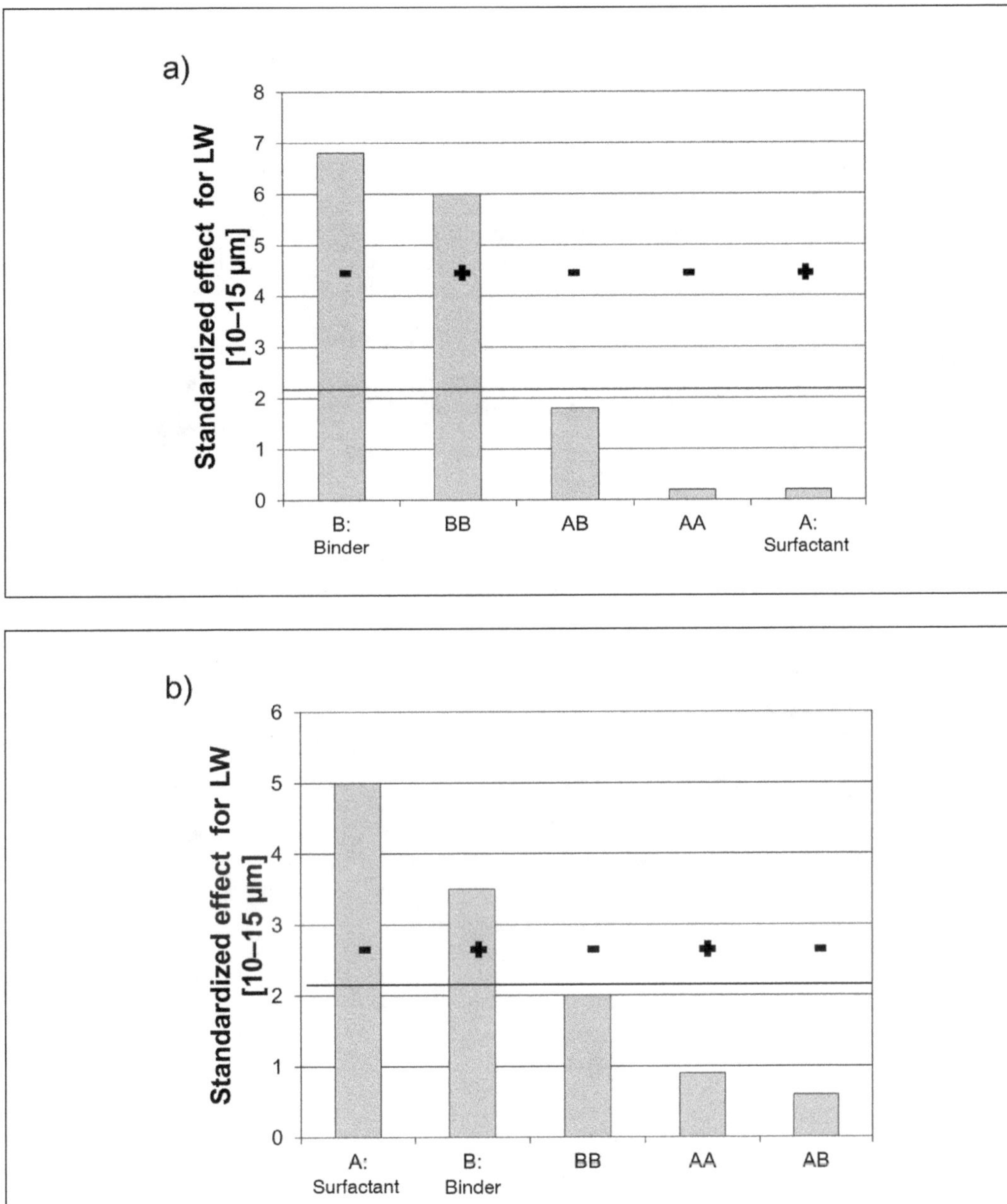

Figure 4.9: Pareto-plot of the new (a) and the old (b) rheology module for the rheological optimization of a base coat

It is obvious from Figure 4.9a that the binder is the dominant factor of influence in case of the new rheology module. Thus, the response LW is in the experimental area robust to fluctuations of the parameter solvent content. In case of the old module both, the binder and the solvent content have a strong effect (see Figure 4.9b). This is also obvious in the response surface in Figure 4.10a. In addition, in case of the new rheology module a saddle point (approximately 25 % solvent and 32 % binder) is present.

However, be aware that a negative statement about robustness is not possible just because of the presence of such a saddle point. Crucial are the slopes of the response surface and the

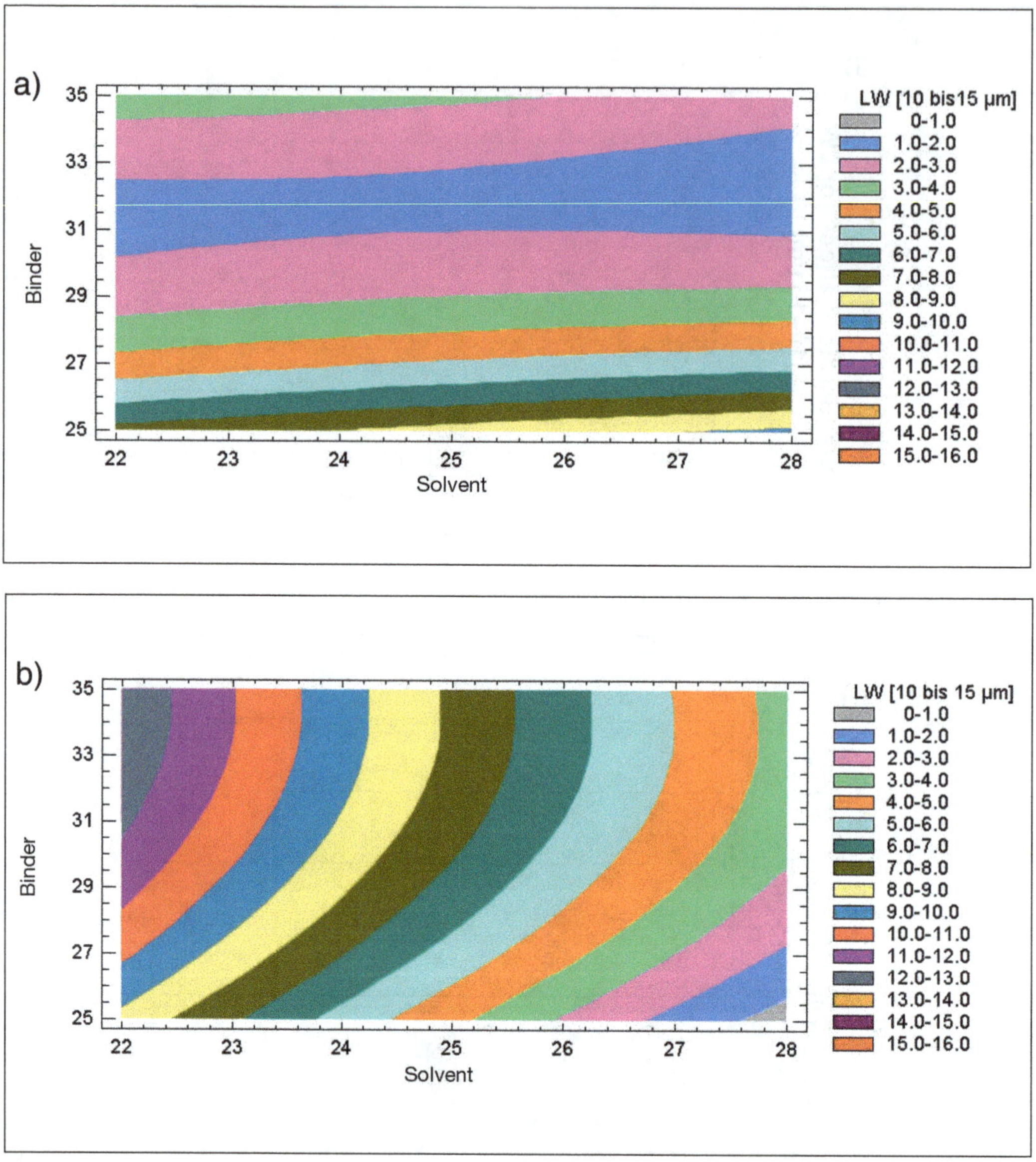

Figure 4.10: Comparison of the contour plot of the new (a) and the old (b) rheology module for the rheological optimization of a base coat

absolute values at this point. Sometimes it is better to work at a saddle point or in a region of saturation, because this is even better than a linear behaviour with high slope.

In this example a minimal value of the response long wave LW should be achieved. Thus, the new module offers clear advantages due to the lower LW-value. In addition, levelling is more or less independent from the solvent content. In a wide range from 29 to 34 % binder only a LW results from 1 up to 3. To achieve such a result with the old rheology module it would be necessary to work in a much small area (binder from 25 to 29 % and solvent from 26 to 28 %). Therefore, robustness studies are quite good possible and common to be done by qualitative analysis of the response surface.

4.3.2 *Quantitative analysis of the regression model*

Response surfaces can be also analyzed in a quantitative manner regarding robustness. Thus an experimental design is generated based on the relevant factors. Usually the parameter selection is oriented on typical factors of influence in case of products or proceeding analytical techniques or the production respective application of products (e.g. temperature, pH-value). The system should be described in an optimal manner by choosing the responses. If limits (e.g. criteria from standards) are known, they have to be included. Afterwards, the experiments are performed and the data analysis takes place. Effects and variances should be small, because parameter variation takes place only in a really small area (working close to reproducible conditions). Regarding the regression model and the specifications of the response values 4 different cases have to be distinguished:

- All responses are inside the specifications, the model is not significant – this is the ideal case. No influence factor affects the response significantly!
- All responses are inside the specifications, but the model is significant. E.g. one influence factor is weak significant. In this situation it is mostly helpful to limit this parameter to a certain region, and keep the effect below the level of significance.
- All responses are outside the specifications, the model is not significant. This case happens very often with outlier or worse goodness of fit.
- All responses are outside the specifications, the model is not significant. This case happens very often with outlier and/or worse goodness of fit.
- All responses are outside the specifications and the model is significant. In this case only reducing the possible region of the response values end up in a robust result.

4.3.3 *Taguchi method*

This method has been developed by the Japanese *Taguchi*, which is a co-founder of the success story of the Japanese car industry [15, 17, 18, 21, 29]. At this time *Taguchi* focused a completely new way: primarily minimization of variance is not achieved by very small tolerances, because this is correlated with high costs. Small variance has to be achieved by robust processes and products! Thus, noise factors and process parameters should have less influence on the product properties. *Taguchi's* radical insight was that the exact choice of values required is under-specified by the performance requirements of the system. In many circumstances, this allows the parameters to be chosen so as to minimize the effects on performance arising from variation in manufacture, environment and cumulative damage (= robustification). Thus, *Taguchi* designs consist of an inner array for all control factors (easy to control) and an outer array including all noise factors (hard to control). They seek to exploit relationships and optimize settings that minimize the effects of the noise variables. First of all, influence parameters on the variance are optimized and afterwards the parameters are adjusted to fulfil the requirements. *Taguchi* argues that it is much easier to adjust means than variances. The function, which has to be optimized, is the ratio of signal to noise defined by Equation 4.3:

Equation 4.3: $$\frac{S}{N} = 10 * log\left(\frac{\bar{y}^2}{s^2}\right)$$

with

$\bar{y}$ = mean

s = standard deviation

An increasing S/N indicates a growing robustness.

Table 4.8: Factor settings regarding the optimization of the micro foam in a thick-coat glaze finish

Factors	Low limit	High limit	Steps
Additive 1	0.1	0.3	2
Additive 2	0.7	1.0	2
Air temperatur	10	30	2
Air humidity	50	90	2

Data analysis is done in 4 steps:

- Search for the influence factors with the largest S/N-ratio
- Search for the response with the largest S/N-ratio
- Define the parameter setting regarding the experiment with the largest S/N-ratio of the responses
- Adjust the mean with the still remaining free controllable factors

▶ ***Basic principle:*** It is much easier to adjust means than variances. Thus, the *Taguchi*-method has the objective to find conditions where simultaneously the responses have a low variability and values close to target.

Example: Micro foam in a thick-coat glaze finish

A thick-coat glaze finish contains two defoaming additives. The aim is to find a composition which is robust against the micro foam regarding ÖNORM B 3803. In addition, the number of bubbles should be minimized. Noise factors are the air humidity and the temperature during the application.

Properties of the not randomized experimental design (see Table 4.9):

- Number of control factors: 2 (see Table 4.8)
- Number of noise factors: 2 (see Table 4.8)
- Number of blocks: 1
- Number of target values: 1
- Number of experiments: 16

Table 4.9: Design matrix, response values and analysis by the Taguchi-method

			Noise factors						
	Iner array control factors		-	+	-	+	Air humidity		
			-	-	+	+	Air temperature		
	Additive 1	Additive 2	Number of bubbles/cm – micro foam				$\bar{y}$	s	S/N
(1)	-	-	28	22	30	22	25.5	4.12	15.83
a	+	-	15	10	19	17	15.25	3.86	11.93
b	-	+	10.5	11	12	11	11.13	0.63	24.95
ab	+	+	8	7	12	9	9	2.16	12.39
HE($\bar{y}$)	-6.19	-10.31							
HE(s)	0.64	-2.60							
HE(S/N)	-8.23	4.80							

dered with another type of illustration, the so called **cumulative distribution curve**. The cumulative frequency can be calculated by summation of the frequencies from data point to data point (see Table A3.1). Based on these data, the so called cumulative frequency can be derived, which is also illustrated versus the measured data (see Figure A3.4b). Very often, data are analyzed in classes like already described in Appendix 2. In this case, the cumulative frequency can be calculated by summation of the frequencies from class to class.

The **relative cumulative frequency** on the interface between two classes or the border between two different data is equal to the so called quantils.

This type of illustration is possible for samples as well as population. The quantils of the normal distribution can be determined from statistic tables or also approximation procedures. Figure A3.3 illustrates the topic.

The bell-shape of the normal distribution can be graphically illustrated as straight line with so called **normal probability plots**. This meets the demand for linear plots, but also certain data sets can be analyzed very easy for their fit with the normal distribution. This kind of data analysis is very often used in case of residuals (see Chapter 3.3). But how can it be realized? The normal probability plot is derived from the cumulative curve by drawing the quantils of the normal distribution with equal distance on the abscissa versus the accompanying quantils of the experimental data set with equal distance on the ordinate. Very often also the relative cumulative frequency is drawn versus the experimental data. These diagrams are also called **probability plots** or **quantil-quantil plots** or **Q-Q-plots**. Statistical software derives these plots automatically, but in case of the probability plot also forms are available like logarithmic forms or millimetre paper ruled in millimetre squares.

If the data points are closely on a straight line, normal distribution is confirmed. Single data points which deviate from the line might be outliers. This can be evaluated by special methods (see Appendix 5 for general remarks. Otherwise it's referred to the special literature). The ideal straight line is very often calculated only between the 25 %- and the 75 %-quantil. From the probability plot it is possible to derive directly for x = 0 the arithmetic mean and the value of 2 * standard deviation between x = -1 and x = +1.

Example: pH-value of a lime paint

The values from Table A2.1 can also be analyzed with a frequency table and a distribution curve (see Figure A3.4a). In this case data are arranged in classes and the frequency is also calculated based on the classes (see Table A3.1). The cumulative frequency can be calculated

Table A3.1: Data and frequency table of the pH-measurement in case of a lime paint

Class in pH units	Absolute frequency	Absolute cumulative frequency	Relative cumulative frequency in [%]	Accompanying quantil of the normal distribution
10 bis 11	1	1	2.63	-1.598
11 bis 12	3	4	18.42	-0.803
12 bis 13	4	8	39.47	-0.199
13 bis 14	6	14	71.05	0.634
14 bis 15	2	16	81.58	1.003
15 bis 16	2	18	92.11	1.616
16 bis 17	1	19	97.37	-
	19			

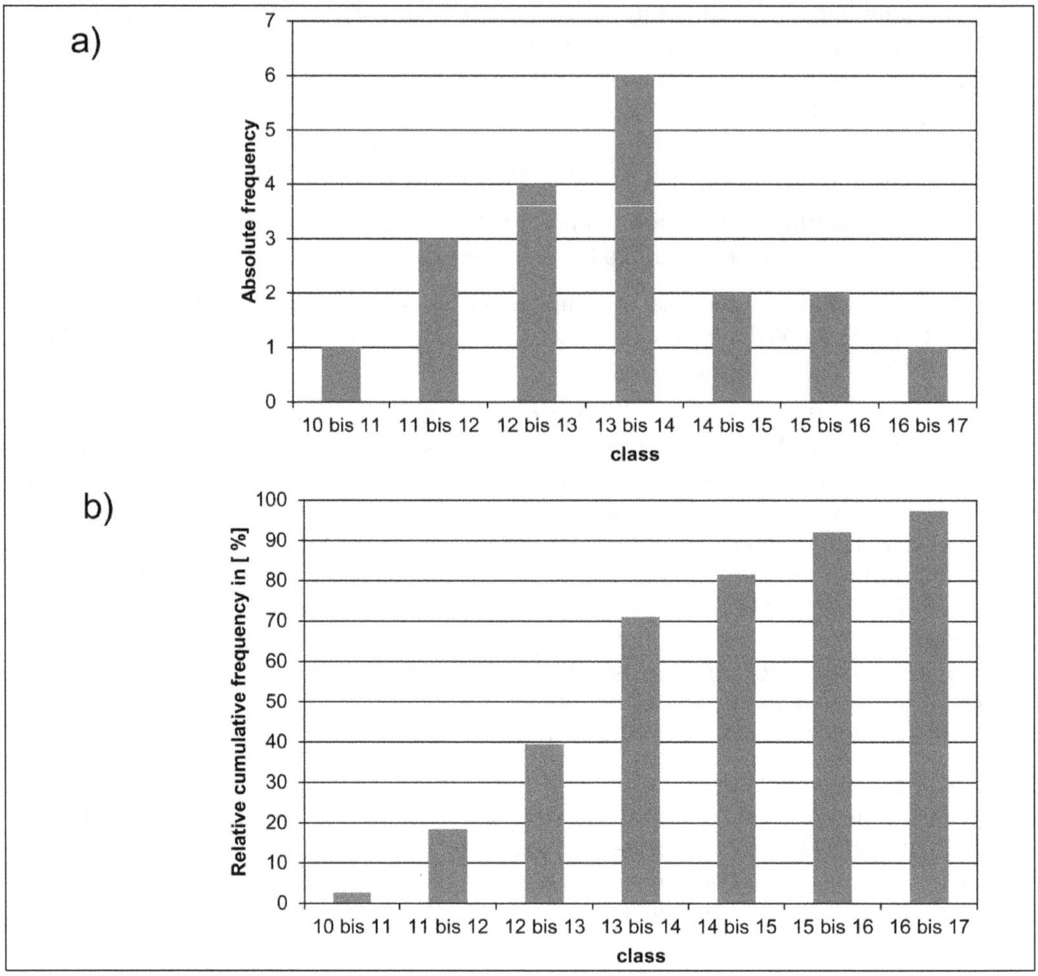

Figure A3.4: a) Absolute frequency distribution; b) Relative cumulative distribution curve

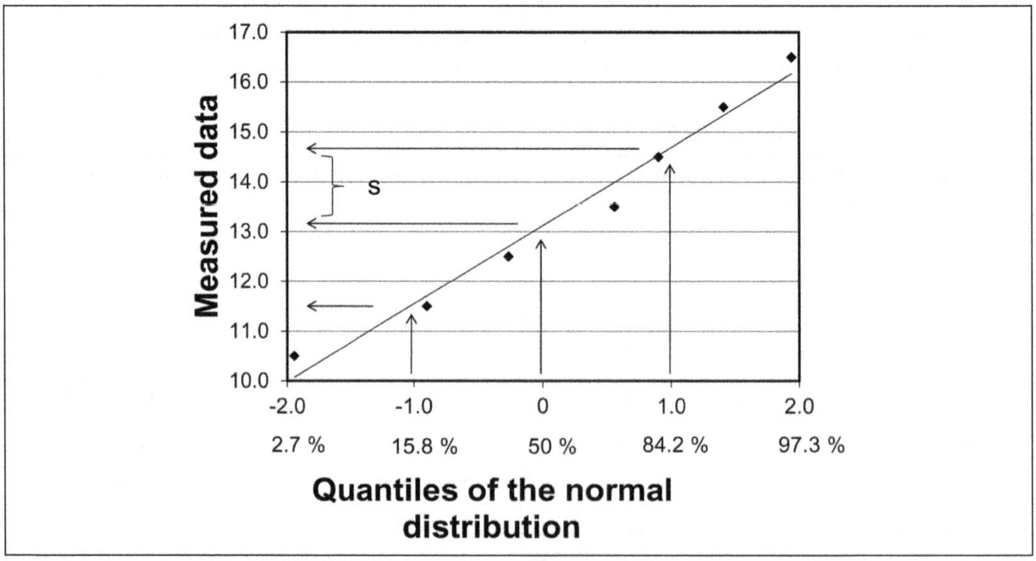

Figure A3.5: Normal probability plot

by summation of the frequencies from data point to data point or class to class (see Table A3.1). Based on this also the cumulative curve can be used to illustrate the data (see Figure A3.4b). In addition, the normal probability plot can be drawn (see Figure A3.5). The data are arranged close to a straight line. Thus, it is possible to assume a normal distribution of the data.

References

- J. Hartung, B. Elpelt, K.-H. Klösener, Statistik, München, 2002
- J. M. Chambers, W. S. Cleveland, Beat Kleiner, Paul A. Tukey: Graphical Methods for Data Analysis, Wadsworth, 1983
- L. Fahrmeier, R. Künstler, I. Pigeot, G. Trutz, Statistik, Der Weg zur Datenanalyse, 5. Auflage, Springer, Berlin, 2006
- J. Hartung, Statistische Tests und Kennzahlen, Statistik. Lehr- und Handbuch der angewandten Statistik. Oldenbourg Verlag, 1991

Appendix 4 - Confidence intervals

The exact value of the mean for a population of data - so called true value - can never be determined exactly because a determination requires infinite number of measurements. However, statistics allows us to set limits around an experimentally determined mean $\bar{x}$ and the true value μ lies within these limits with a given degree of probability (**confidence level** α). These limits are called **confidence limits**, and the interval they define is known as the **confidence interval**. However, to a certain extend the confidence interval does not include the true value. This probability is 100 % - α.

Confidence intervals offer the great advantage compaired to just single data points, that the significance of a statement is already included. If the size of the interval is too large with a given degree of confidence, maybe the sample size out of the data population is too small, or the examined phenomenon has a too low precision. Thus, only the measurement of a huge sample could end up in an acceptable size of the confidence interval.

The standard deviation s is a good estimation of σ in case of a great data set. Often we are faced with limitation in time or costs prevent us from an accurate estimation of σ. However, if s is calculated from a small set of data it is a quite uncertain estimation of σ. Thus, the confidence limits are necessarily broader and the so-called student distribution describes the set of data much better than the Gaussian distribution. As correction factor between the distribution which describes the sample out of a population and the Gaussian distribution for the population the important statistical parameter t is used. It can be derived from the Student distribution and depends on the confidence level (see Table A4.1). Most people will feel that the more experiments we make, the more reliable is the estimation of μ. To allow this, statistics modify the Equation A4.1 by dividing the standard deviation with the square root of the number of experiments. The **confidence interval for the mean** $\bar{x}$ is defined by Equation A4.1:

Equation A4.1: $$\bar{x} \pm \frac{t * s}{\sqrt{N}}$$

t = Student Factor (depending on the number of replicate measurements N and the level of confidence 1-α) as correction factor for the normal distribution

N = number of replicate measurements

The confidence interval of a mean expresses the accuracy of a mean from a small data set. The interval decrease with an increasing number of replicates. However, this interval is not correlated with the interval of a single measurement. This is calculated directly from the products of s*t. In case of single measurements the terminus **tolerance interval** is very often used. There is a difference, if 50 measurements are performed, or the mean is calculated from 5 means of 5 experimental series each based on 10 replicates. These 5 means show a much more narrow distribution then 50 single values of a whole data set. The means are also scattered more or less symmetrically around the true value and represent a sample out of the population of the means. This population is different from that, which describes a sample of 50 single replicates.

The confidence limits describe an area below the normal distribution curve of single values or means, which is different from the value directly calculated from the confidence level α (e.g. 86 % instead of 95 %). Also the probability is a variable with certain fluctuations and can

Table A4.1: Values of t for various levels of probability to calculate the confidence interval (for a two-tailed test, see Appendix 5)

Degrees of freedom (= number n of measurements -1)	t for the confidence level (1-α)		
	95 %	99 %	99.9 %
1	12.71	63.66	636.62
2	4.30	9.92	31.60
3	3.18	5.82	12.92
4	2.78	4.60	8.61
5	2.57	4.03	6.87
6	2.36	3.71	5.96
7	2.31	3.36	5.41
8	2.26	3.25	5.04
9	2.26	3.25	4.78
10	2.22	3.17	4.56
18	2.10	2.88	3.92
20	2.09	2.86	3.85
100	1.98	2.63	3.39
> 200	1.96	2.58	3.29
Probability in case of a one-tailed test	97.5 %	99.5 %	99.95 %

only be estimated by the factor t. In case of a lot of samples, the value is on average at 95 %. However, as long as we deal with samples from the population, only estimations are possible.

There is a clear consequence from these facts: The greater the certainty, the greater the interval required - precise statements are never valid and valid statements are always not

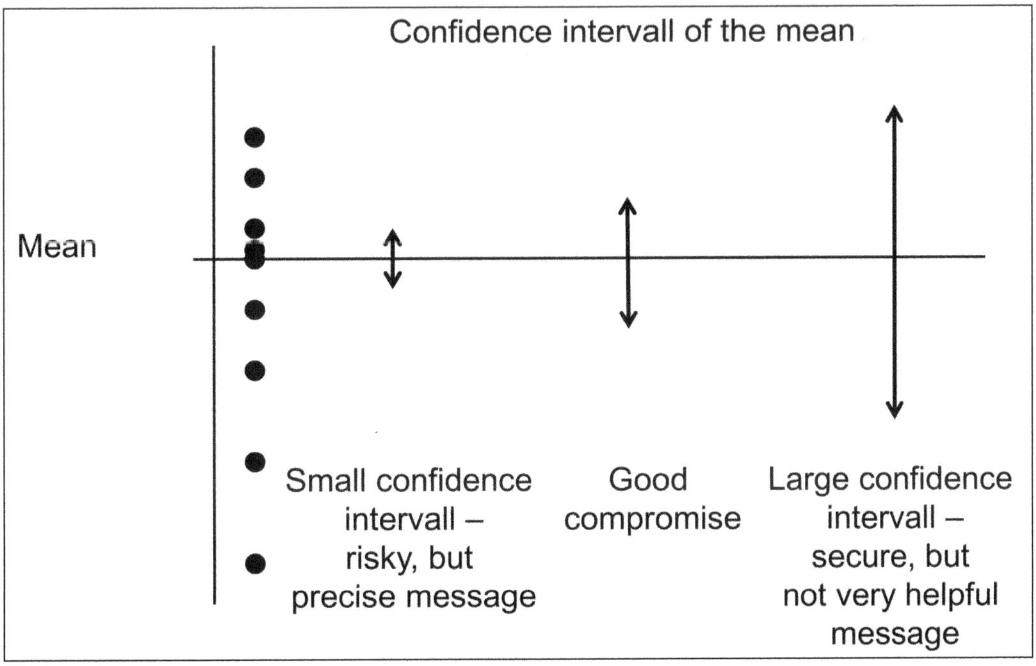

Figure A4.1: Effect of different confidence intervals on the sense of statements

precise. A low probability (e.g. 80 % instead of 95 %) reduces the confidence interval, but increases the risk, that the true value is not inside this interval. Confidence intervals with high probability (e.g. 99 %) are valid, but due to the huge interval not precise (Figure A4.1).

▶ ***Basic principle:*** The greater the confidence interval, the greater the certainty that the true value is inside this interval. Statements will be much more secure, but lose their explanatory power, because the definition is diffuse. In principle it is impossible to make absolute statements by statistics. Statements will be defined and provided with a certain amount of reliability.

Example: pH-value of lime paints – continuation

Values are used from Table A2.1 and Appendix 2. The confidence interval is calculated for 95 and 99 % probability.

$$95\,\%:\ \mu = \bar{x} \pm \frac{t * s}{\sqrt{N}} = 13.29 \pm \frac{2.10 * 1.55}{\sqrt{19}} = 13.29 \pm 0.75 \text{ or } VB(95\,\%, f = 18) = 0.75$$

$$99\,\%:\ \mu = \bar{x} \pm \frac{t * s}{\sqrt{N}} = 13.29 \pm \frac{2.88 * 1.55}{\sqrt{19}} = 13.29 \pm 1.02 \text{ or } VB(95\,\%, f = 18) = 1.02$$

If differences are analyzed as in DoE usually the case, the confidence interval is calculated by Equation A4.2:

Equation A4.2: $$\bar{x} \pm \frac{2 * t * s}{\sqrt{N}}$$

N is the number of measurements for both average values

In Equation A4.2 an additional factor 2 is present, based on the fact that the standard deviation of a difference of two average values is twice the standard deviation of one average value. This is also illustrated in Figure 1.21. The difference between two means can be increased or decreased only by two times the standard deviation.

Very often the so-called standard deviation of the effects is used (see Chapter 1.6).

Equation A4.3: $$s_{Effekt} = \frac{2 * s}{\sqrt{N}}$$

The degree of freedom is reduced to f = N - 2.

Author

Dr. Dipl. Ing. Albert Rössler studied chemical engineering at the engineering college HTL Wels, the Royal Institute of Technology Stockholm and the ETH Zurich with a PhD at ETH Zurich. Since 2003 he is working at one of the European competency leader in wood preservatives and paints ADLER-Werk Lackfabrik Johann Berghofer GmbH & Co KG, Austria in the R&D department. Since 2009 he is head of this section. In addition, since 2012 he is lecturer at the University Innsbruck and the Management Center Innsbruck. He is author of several patents and published a great number of scientific papers in the field of coatings, fuel cells, color and textile chemistry, chemical reaction engineering and nanoparticle technology by flame spray pyrolysis.

Index

Symbols

A

B

C

D

E

F

G

H

I

L

M

N

O

P

Q

R

S

T

U

V